Advances in Intelligent

Volume 1091

Series Editor

Janusz Kacprzyk, Systems Research Institute, Polish Academy of Sciences, Warsaw, Poland

Advisory Editors

Nikhil R. Pal, Indian Statistical Institute, Kolkata, India
Rafael Bello Perez, Faculty of Mathematics, Physics and Computing, Universidad Central de Las Villas, Santa Clara, Cuba
Emilio S. Corchado, University of Salamanca, Salamanca, Spain
Hani Hagras, School of Computer Science and Electronic Engineering, University of Essex, Colchester, UK
László T. Kóczy, Department of Automation, Széchenyi István University, Gyor, Hungary
Vladik Kreinovich, Department of Computer Science, University of Texas at El Paso, El Paso, TX, USA
Chin-Teng Lin, Department of Electrical Engineering, National Chiao Tung University, Hsinchu, Taiwan
Jie Lu, Faculty of Engineering and Information Technology, University of Technology Sydney, Sydney, NSW, Australia
Patricia Melin, Graduate Program of Computer Science, Tijuana Institute of Technology, Tijuana, Mexico
Nadia Nedjah, Department of Electronics Engineering, University of Rio de Janeiro, Rio de Janeiro, Brazil
Ngoc Thanh Nguyen, Faculty of Computer Science and Management, Wrocław University of Technology, Wrocław, Poland
Jun Wang, Department of Mechanical and Automation Engineering, The Chinese University of Hong Kong, Shatin, Hong Kong

The series "Advances in Intelligent Systems and Computing" contains publications on theory, applications, and design methods of Intelligent Systems and Intelligent Computing. Virtually all disciplines such as engineering, natural sciences, computer and information science, ICT, economics, business, e-commerce, environment, healthcare, life science are covered. The list of topics spans all the areas of modern intelligent systems and computing such as: computational intelligence, soft computing including neural networks, fuzzy systems, evolutionary computing and the fusion of these paradigms, social intelligence, ambient intelligence, computational neuroscience, artificial life, virtual worlds and society, cognitive science and systems, Perception and Vision, DNA and immune based systems, self-organizing and adaptive systems, e-Learning and teaching, human-centered and human-centric computing, recommender systems, intelligent control, robotics and mechatronics including human-machine teaming, knowledge-based paradigms, learning paradigms, machine ethics, intelligent data analysis, knowledge management, intelligent agents, intelligent decision making and support, intelligent network security, trust management, interactive entertainment, Web intelligence and multimedia.

The publications within "Advances in Intelligent Systems and Computing" are primarily proceedings of important conferences, symposia and congresses. They cover significant recent developments in the field, both of a foundational and applicable character. An important characteristic feature of the series is the short publication time and world-wide distribution. This permits a rapid and broad dissemination of research results.

** **Indexing: The books of this series are submitted to ISI Proceedings, EI-Compendex, DBLP, SCOPUS, Google Scholar and Springerlink** **

More information about this series at http://www.springer.com/series/11156

Grzegorz Sierpiński
Editor

Smart and Green Solutions for Transport Systems

16th Scientific and Technical Conference
"Transport Systems. Theory and Practice 2019"
Selected Papers

Editor
Grzegorz Sierpiński
Faculty of Transport
Silesian University of Technology
Katowice, Poland

ISSN 2194-5357 ISSN 2194-5365 (electronic)
Advances in Intelligent Systems and Computing
ISBN 978-3-030-35542-5 ISBN 978-3-030-35543-2 (eBook)
https://doi.org/10.1007/978-3-030-35543-2

© Springer Nature Switzerland AG 2020
This work is subject to copyright. All rights are reserved by the Publisher, whether the whole or part of the material is concerned, specifically the rights of translation, reprinting, reuse of illustrations, recitation, broadcasting, reproduction on microfilms or in any other physical way, and transmission or information storage and retrieval, electronic adaptation, computer software, or by similar or dissimilar methodology now known or hereafter developed.
The use of general descriptive names, registered names, trademarks, service marks, etc. in this publication does not imply, even in the absence of a specific statement, that such names are exempt from the relevant protective laws and regulations and therefore free for general use.
The publisher, the authors and the editors are safe to assume that the advice and information in this book are believed to be true and accurate at the date of publication. Neither the publisher nor the authors or the editors give a warranty, expressed or implied, with respect to the material contained herein or for any errors or omissions that may have been made. The publisher remains neutral with regard to jurisdictional claims in published maps and institutional affiliations.

This Springer imprint is published by the registered company Springer Nature Switzerland AG
The registered company address is: Gewerbestrasse 11, 6330 Cham, Switzerland

Preface

Smart is a notion which relates to using solutions based on advanced technologies. In smart cities, information performs important and diversified functions. Among many others, it enables transport of people and cargo, while improving its efficiency and limiting its negative environmental impact. On the other hand, collecting data about travelling should guarantee that the needs of the travelling population are properly determined. It is for the knowledge concerning this area that transport can be performed according to the principles of sustainable development, meaning a combination of reasonable use of energy and technological development making it possible to transform urban areas into smart and green cities.

This publication contains selected papers submitted to and presented at the 16th "Transport Systems. Theory and Practice" Scientific and Technical Conference organized by the Department of Transport Systems and Traffic Engineering, Faculty of Transport, Silesian University of Technology, Katowice, Poland. The problems addressed in the publication entitled *Smart and Green Solutions for Transport Systems* have been divided into four parts:

- Part 1. Smart Cities Components in the Service of Sustainable Transport,
- Part 2. Optimization as a Way to Improve the Efficiency of Transport Systems,
- Part 3. Tools Supporting the Reduction of Negative Environmental Impact in Transport,
- Part 4. Advanced Methods for Data Collection and Analysis.

The articles included in the publication are expressions of case study-based scientific and practical approach to the problems of contemporary transport systems. The optimization methods and models proposed to date as well as the data gathered using advanced methods provide grounds for a system approach to the assessment of the solutions currently at hand. On the other hand, the perspective of spatial planning and application of the Intelligent Transport Systems proposed in the publication may facilitate the development of truly smart cities while emphasizing the need for reduction of the negative environmental impact of transport.

I would like to express my deepest gratitude to all authors, for reflecting the key problems of contemporary transport systems in a concise manner, as well as to reviewers, in recognition of their insightful remarks and suggestions without which this collection of papers would have never been published.

September 2019   Grzegorz Sierpiński

Organization

The 16th "Transport Systems. Theory and Practice" (TSTP2019) Scientific and Technical Conference is organized by the Department of Transport Systems and Traffic Engineering, Faculty of Transport, Silesian University of Technology, Poland.

Organizing Committee

Organizing Chair

Grzegorz Sierpiński   Silesian University of Technology, Poland

Members

Marcin Staniek
Renata Żochowska
Ireneusz Celiński
Grzegorz Karoń
Marcin J. Kłos
Krzysztof Krawiec
Aleksander Sobota
Barbara Borówka
Piotr Soczówka

The Conference Took Place Under the Honorary Patronage

Ministry of Infrastructure
Marshal of the Silesian Voivodeship

Scientific Committee

Stanisław Krawiec (Chairman)	Silesian University of Technology, Poland
Rahmi Akçelik	SIDRA SOLUTIONS, Australia
Tomasz Ambroziak	Warsaw University of Technology, Poland
Henryk Bałuch	The Railway Institute, Poland
Werner Brilon	Ruhr-University Bochum, Germany
Margarida Coelho	University of Aveiro, Portugal
Boris Davydov	Far Eastern State Transport University, Khabarovsk, Russia
Mehmet Dikmen	Baskent University, Turkey
Domokos Esztergár-Kiss	Budapest University of Technology and Economics, Hungary
Zoltán Fazekas	Institute for Computer Science and Control, Hungary
József Gál	University of Szeged, Hungary
Anna Granà	University of Palermo, Italy
Andrzej S. Grzelakowski	Gdynia Maritime University, Poland
Mehmet Serdar Güzel	Ankara University, Turkey
József Hansel	AGH University of Science and Technology Cracow, Poland
Libor Ižvolt	University of Žilina, Slovakia
Marianna Jacyna	Warsaw University of Technology, Poland
Ilona Jacyna-Gołda	Warsaw University of Technology, Poland
Nan Kang	Tokyo University of Science, Japan
Jan Kempa	University of Technology and Life Sciences in Bydgoszcz, Poland
Michael Koniordos	Piraeus University of Applied Sciences, Greece
Bogusław Łazarz	Silesian University of Technology, Poland
Michal Maciejewski	Technical University Berlin, Germany
Elżbieta Macioszek	Silesian University of Technology, Poland
Ján Mandula	Technical University of Košice, Slovakia
Sylwester Markusik	Silesian University of Technology, Poland
Antonio Masegosa	IKERBASQUE Research Fellow at University of Deusto Bilbao, Spain
Agnieszka Merkisz-Guranowska	Poznań University of Technology, Poland
Anna Mężyk	Kazimierz Pulaski University of Technology and Humanities in Radom, Poland
Maria Michałowska	University of Economics in Katowice, Poland
Leszek Mindur	International School of Logistic and Transport in Wrocław, Poland
Maciej Mindur	Lublin University of Technology, Poland
Goran Mladenović	University of Belgrade, Serbia

Kai Nagel	Technical University Berlin, Germany
Piotr Niedzielski	University of Szczecin, Poland
Piotr Olszewski	Warsaw University of Technology, Poland
Enrique Onieva	Deusto Institute of Technology University of Deusto Bilbao, Spain
Asier Perallos	Deusto Institute of Technology University of Deusto Bilbao, Spain
Hrvoje Pilko	University of Zagreb, Croatia
Antonio Pratelli	University of Pisa, Italy
Dariusz Pyza	Warsaw University of Technology, Poland
Cesar Queiroz	World Bank Consultant (Former World Bank Highways Adviser), Washington, DC, USA
Andrzej Rudnicki	Cracow University of Technology, Poland
František Schlosser	University of Žilina, Slovakia
Grzegorz Sierpiński	Silesian University of Technology, Poland
Jacek Skorupski	Warsaw University of Technology, Poland
Aleksander Sładkowski	Silesian University of Technology, Poland
Wiesław Starowicz	Cracow University of Technology, Poland
Andrzej Szarata	Cracow University of Technology, Poland
Tomasz Szczuraszek	University of Technology and Life Sciences in Bydgoszcz, Poland
Antoni Szydło	Wrocław University of Technology, Poland
Grzegorz Ślaski	Poznań University of Technology, Poland
Paweł Śniady	Wrocław University of Environmental and Life Sciences, Poland
Andrew P. Tarko	Purdue University West Lafayette, USA
Frane Urem	Polytechnic of Šibenik, Croatia
Hua-lan Wang	School of Traffic and Transportation, Lanzhou Jiaotong University, Lanzhou, China
Mariusz Wasiak	Warsaw University of Technology, Poland
Adam Weintrit	Gdynia Maritime University, Poland
Andrzej Więckowski	AGH University of Science and Technology Cracow, Poland
Katarzyna Węgrzyn - Wolska	Engineering School of Digital Science Villejuif, France
Adam Wolski	Polish Naval Academy, Gdynia, Poland
Olgierd Wyszomirski	University of Gdańsk, Poland
Elżbieta Załoga	University of Szczecin, Poland
Stanisława Zamkowska	Blessed Priest Władysław Findysz University of Podkarpacie in Jaslo, Poland
Jacek Żak	Poznań University of Technology, Poland
Jolanta Żak	Warsaw University of Technology, Poland

Referees

Rahmi Akçelik
Przemysław Borkowski
Tony Castillo-Calzadilla
Ireneusz Celiński
Jacek Chmielewski
Piotr Czech
Magdalena Dobiszewska
Michal Fabian
Barbara Galińska
Anna Granà
Róbert Grega
Mehmet Serdar Güzel
Katarzyna Hebel
Peter Jenček
Nan Kang
Peter Kaššay
Jozef Kuĺka
Michał Maciejewski
Elżbieta Macioszek
Krzysztof Małecki
Martin Mantič
Paola di Mascio
Silvia Medvecká-Beňová
Maria Mrówczyńska
Vitalii Naumov
Katarzyna Nosal Hoy
Ander Pijoan
Hrvoje Pilko
Michal Puškár
Alžbeta Sapietová
Grzegorz Sierpiński
Izabela Skrzypczak
Marcin Staniek
Dariusz Tłoczyński
Andrzej Wieckowski
Grzegorz Wojnar
Adam Wolski
Ninoslav Zuber

Contents

Smart Cities Components in the Service of Sustainable Transport

Planning Spatial Development of a City from the Perspective of Its Residents' Mobility Needs 3
Tomasz Szczuraszek and Jacek Chmielewski

The Impact of Cities' Spatial Planning on the Development of a Sustainable Urban Transport 13
Grzegorz Dydkowski

The Role of Intelligent Transport Systems in the Development of the Idea of Smart City 26
Wojciech Lewicki, Bogusław Stankiewicz, and Aleksandra A. Olejarz-Wahba

How to Level the Playing Field for Ride-Hailing and Taxis 37
Kaan Yıldızgöz and Hüseyin Murat Çelik

Assessment of BEVs Use in Urban Environment 52
Ewelina Sendek-Matysiak

Optimization as a Way to Improve the Efficiency of Transport Systems

MCDM as the Tool of Intelligent Decision Making in Transport. Case Study Analysis 67
Barbara Galińska

AHP as a Method Supporting the Decision-Making Process in the Choice of Road Building Technology 80
Izabela Skrzypczak, Wanda Kokoszka, Tomasz Pytlowany, and Wojciech Radwański

**Multiple Criteria Evaluation of the Planned Bikesharing System
in Jaworzno** 93
Andrzej Bąk, Katarzyna Nosal Hoy, and Katarzyna Solecka

**The Application of Bus Rapid Transit System in the City
of Baquba and Its Impact on Reducing Daily Trips** 104
Firas Alrawi and Yagoob Hadi

**The Application of Lithium-Ion Batteries for Power Supply
of Railway Passenger Cars and Key Approaches
for System Development** 114
Viacheslav Bondarenko, Dmytro Skurikhin, and Jerzy Wojciechowski

**Time Parameters Optimization of the Export Grain Traffic
in the Port Railway Transport Technology System** 126
Oleg Chislov, Taras Bogachev, Alexandra Kravets, Victor Bogachev,
Vyacheslav Zadorozhniy, and Irina Egorova

**Tools Supporting the Reduction of Negative Environmental
Impact in Transport**

**Strategic Planning of the Development of Trolleybus
Transportation Within the Cities of Poland** 141
Marcin Wołek and Katarzyna Hebel

**Ways to Improve Sustainability of the City Transport System
in the Municipal Gas-Engine Vehicles' Fleet Growth** 153
Irina Makarova, Ksenia Shubenkova, Larisa Gabsalikhova,
Gulnaz Sadygova, and Eduard Mukhametdinov

Electric Vehicles - Problems and Issues 169
Elżbieta Macioszek

**Estimating Pollutant Emissions Based on Speed Profiles
at Urban Roundabouts: A Pilot Study** 184
Orazio Giuffrè, Anna Granà, Tullio Giuffrè, Francesco Acuto,
and Maria Luisa Tumminello

**Examining the Influence of Railway Track Routing on the Thermal
Regime of the Track Substructure – Experimental Monitoring** 201
Peter Dobeš, Libor Ižvolt, and Stanislav Hodás

Advanced Methods for Data Collection and Analysis

**Estimating Parameters of Demand for Trips by Public Bicycle
System Using GPS Data** 213
Vitalii Naumov and Krystian Banet

Support for Pro-ecological Solutions in Smart Cities with the Use of Travel Databases – a Case Study Based on a Bike-Sharing System in Budapest 225
Katarzyna Turoń, Grzegorz Sierpiński, and János Tóth

Finding the Way at Kraków Główny Railway Station: Preliminary Eye Tracker Experiment 238
Anton Pashkevich, Eduard Bairamov, Tomasz E. Burghardt, and Matus Sucha

Using the Kalman Filter for Purposes of Road Condition Assessment 254
Marcin Staniek

Traffic Measurements for Development a Transport Model 265
Marcin Jacek Kłos, Aleksander Sobota, Renata Żochowska, and Piotr Soczówka

Life Cycle Sustainability Assessment of Sport Utility Vehicles: The Case for Qatar 279
Nour N. M. Aboushaqrah, Nuri Cihat Onat, Murat Kucukvar, and Rateb Jabbar

Author Index 289

Smart Cities Components in the Service of Sustainable Transport

Planning Spatial Development of a City from the Perspective of Its Residents' Mobility Needs

Tomasz Szczuraszek and Jacek Chmielewski[✉]

University of Science and Technology, Bydgoszcz, Poland
{tomasz.szczuraszek,jacek-ch}@utp.edu.pl

Abstract. This paper provides an analysis of the influence of an urban spatial planning policy on the living standards and the costs of living of the city's residents and on the costs borne by the municipality. In the analysis, two scenarios for the development of spatial planning of a selected city (of approx. 350,000 residents and 176 km^2) were compared: the so-called 'compact city' and 'diffuse city'. A city transport demand model was also used in the analysis. It was developed by the authors for different time horizons of the forecast period. The key objective of the housing policy described in this paper should be to manage the spatial development of the city in the most beneficial way. This requires a thorough multiple-criteria assessment of a number of development scenarios using state-of-the-art IT tools, such as 'transport demand models' used by the authors, and the results of market research into the residents' preferences.

Keywords: Transport policy · Traffic volume · Environment · Transport demand models

1 Introduction

One of the major problems essentially all medium-sized and large cities in Poland and other developed countries have to tackle now [1–7] is the increasing traffic congestion which goes beyond the so-called rush hours [8, 9]. The condition of oversaturation of city road network with vehicular traffic more and more often extends over almost the whole day. As a result, residents of cities affected by problems like this experience their negative consequences: journeys take longer and longer and their costs are higher and higher; the residents come under a growing stress and face an increasing risk of becoming the victims of road accidents as their exposure to excess traffic is longer. On the other hand, from the point of view of city authorities, residents experience the problems of environmental pollution, increased expenditure on the maintenance of public transports and the need to allot new land to expand the road infrastructure [10]. In this way, we deplete the land which could be used by residents to improve the quality of life, for example by creating new recreational areas or green spaces. A similar issue was dealt with by the authors of the paper [11], where the city compactness is pointed as the principal sustainable urban form to combat increasing urbanisation pressures and urban sprawl.

Of course, there are a lot of reasons behind the problem of increasing traffic congestion, from a greater need for mobility, growing numbers of car owners and increasing use of their private cars, an increasing demand for goods transportation, to the aspects of urban spatial planning. In the latter case, at least two processes should be mentioned which have a significant influence on the amount of traffic within cities: the extension of cities and the so-called 'urban sprawl', understand and the phenomena the spreading of urban developments (such as houses and shopping centers) on undeveloped land near a city. In both cases we talk about an increased road performance within urban road networks and extended journey times. As a consequence, traffic volumes grow, traffic conditions get worse and traffic jams get longer.

Cities – like living organisms – undergo continuous changes, even within the framework of long-established spatial structures. This is due to a number of factors, such as the progress of civilization and economic growth, demographic, economic, political as well as social changes in the city itself and the preferences of its residents as regards the quality of life and their expectations of the public space [12].

The spatial planning policy which has recently been quite often applied to cities in Poland assumes the expansion of their outskirts or undeveloped edge districts. The decisions are often justified by the accessibility of the areas and relatively low prices of the land [13]. Furthermore, the areas are supposed to be more competitive than the areas in adjacent municipalities or communes. The basis for the policy is the desire to stop residents from leaving the city for neighbouring locations, so completely new residential districts are developed with attractive natural qualities. However, the most frequent effects of these measures are just the opposite and include the following:

- Internal migration of the residents from central areas to suburban districts and to the outskirts, along with a successive outflow of people to neighbouring communes or municipalities;
- An increase in the number and duration of non-pedestrian journeys, which contributes to the choice of private cars as the only mean of transport;
- The necessity to substantially develop the technical infrastructure, including costly transport systems;
- A small utilization rate of public transport lines to neighbouring districts, high costs of maintenance and operation of these services (with the lines routed through undeveloped or hardly-developed areas) and increased transportation performance (vehicle-kilometers);
- A considerable increase in the expenditure from the city budget to maintain public transport and to construct and maintain the additional network of roads;
- Further spatial division of the city and its social and functional disintegration;
- A smaller than expected utilization rate of the technical infrastructure in which investments have been made so far;
- An increase in the social costs to be borne by the city residents and investors, related to longer travel times and a higher fuel consumption.

Considering the above effects, a decision to plan the development of a city in that direction seems to be anything but reasonable. The only justification for that policy could be a full reversibility of its effects. Yet implementing such a spatial development policy which drains away lots of financial resources will cause permanent changes to the city space and the changes will not guarantee any improvement of the residents' living standards and may even lead to their decline. Remote districts, separated from the rest of the city, will always suffer from an insufficiency of commercial and retail outlets, educational, cultural and transport facilities. Even if the development density is quite high, the local commercial and service potential will be inadequate to the demand. Therefore, such districts will always lack certain services and experience a shortage of jobs, education and culture, which turns them into monofunctional 'dormitory suburbs'. Scattered origins and destinations, which will move further away from one another as a consequence of the processes of urban structural decentralization, will increase the number and duration of everyday trips, and thus will increase the average cost of living for the residents.

The purpose of this article is to demonstrate that a spatial development policy of diffuse city is not very effective for the residents from the point of view of the comfort of life or for the municipality from the economical point of view (mainly due to the increased costs of public transport and the costs of related investments in the road infrastructure). The article presents an innovative approach to spatial development analysis of the city, taking into account the mobility of its residents. The authors indicate the use of transport demand models as the effective mathematical tool to analyse the impact of spatial planning on the costs and living conditions of residents.

The medium size city of Bydgoszcz in Poland was selected to serve as an example. The analysis was performed as part of the 'Transportation study for the city of Bydgoszcz' and utilized a digital simulation model for the transport system in the city and its vicinities (a transport demand model), as a professional approach to the analysis required the use of appropriate IT tools. The complexity of transportation processes, their substantial variability in time and susceptibility to various factors, such as spatial development, external and internal conditions, arrangement and characteristics of individual transport networks, residents' travel behaviours, including their travel preferences, demographics, traffic conditions within the network of roads, etc. makes it impossible to carry out complex spatial analyses related to transports for such a large area in a reliable way without proper tools. Transport demand models provide the possibility - through simulation calculations - to replicate current or potential future transport processes in the study area (e.g. traffic volumes, passenger flow rates, saturation states etc.) for various scenarios of investment in spatial development and transport infrastructure and for other measures with socio-economic consequences.

The software environment for the development of the model was provided by the German VISUM interactive graphic package of PTV Vision from Karlsruhe [14]. The transport demand model was defined using the Polish coordinate system CS2000, Zone 6. The definition ensures correct functionality of the model as regards compatibility with all map systems, including the OpenStreetMap, which significantly facilitates the development and application of the transport model.

In order to develop the above-mentioned transport demand model, a few hundred thousand data was gathered and coded into databases to describe the city and the neighbouring municipalities and communes, as well as their transport networks in high detail level (including every building). Moreover, a number of transport-related studies (including surveys carried out among the residents of the study area) and analyses concerning key processes which affect the local transport [4, 5].

2 Study Area

The study is located in the city of Bydgoszcz, situated in Central Poland and covers the area of approx. 176 km^2. More than 350,000 residents live there and the population density of the city, which is a measure of its compactness and manner of spatial development, is approx. 2,000 people per km^2, being rather small if compared to other large or medium-sized cities in Europe (Lyon – 10,100, München – 4,531, Prague 2,408). This must be due to the specific use of the city space where more than 9,300 ha (53%) are covered by green areas and agricultural lands which are currently excluded from settlement or investment activity. Historical circumstances and geographical conditions have contributed to the city's belt-shaped spatial arrangement, particularly if the undeveloped areas are disregarded (see Fig. 1). The city used to develop along the rivers of Brda and Vistula, initially the only transportation routes in the area, and as a result of administrative decisions to incorporate the nearby towns and villages to the east and west of the city. The city spans a distance of 22 km from the east to the west, but only about 9 km from the north to the south, including the rest and recreation areas in the north of the city, so the vertical to horizontal axis ratio of the city is 1:2.44.

The multi-family residential function is mainly provided by the eastern part of the city - the Fordon district and the central and adjacent districts (see Fig. 1). As a result, a division of Bydgoszcz into two 'internal cities' can be seen where the central part has a population of over 250,000 and is functionally connected with the 'second city' (Fordon) of 76,000 residents but also with quite modest commercial, service, administrative, cultural, recreational etc. facilities. The two main parts of the city are supplemented by the western suburbs where single-family houses prevail.

The spatial development plans for the city initially provided for the so-called 'extension of the city' [9]. In an attempt to stop the outflow of residents to the neighboring communes a proposition was put forward to extend the city's outskirts and develop its outermost areas, which were to provide similar housing conditions to those found outside the urban zone, i.e. in the adjacent communes. The plan was to considerably increase the housing function through the development of multi-family estate projects, especially in the north of the biggest district, Fordon (Fig. 1). This manner of spatial development was subsequently approved in the official local development plan, and the first steps to start its implementation have already been done.

The master zoning plan for the city indicated that the above-mentioned land was clearly outside the most favourable area for the location of housing and from the point of view of social effects (shown as a circle in Fig. 2). The spatial development and natural and topographic conditions in the north and the south made the encircled districts the most attractive and the fastest-growing in the city.

Therefore, the authors of this paper – based on the initial analyses – suggested certain substantial modifications to the assumptions of the official city development plans. The proposition consisted in the following:

- Use the outermost districts indicated in the spatial development plans along with the area neighbouring on Fordon district from the north as the districts intended for single-family houses only, to be built in plots exceeding 1500 m^2;

- Make a better use of the other areas with a substantial urban potential – the areas surrounding the city centre and the central part of the Fordon district – and regenerate (and intensify) the developments in the central districts.

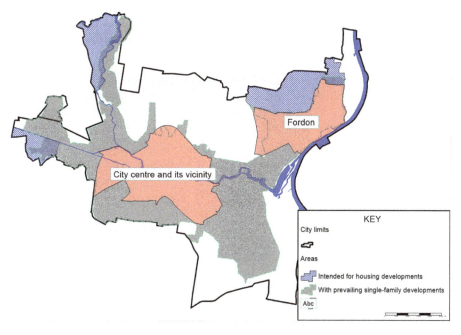

Fig. 1. Map shows planned new residential areas and areas planned for the extension of housing in Bydgoszcz

The authors indicated the areas of enormous urban advantages, whose utilization or promotion is currently marginal. Some of the areas are located both close to the city centre and among leisure and recreational regions. At the same time, it is noted that a better use should be made of certain parts of the oldest estates in the central districts, incorporated into the city from the rural areas in the past. If efficient public private partnership structures were established, the districts, which are now low standard and with a low rate of utilization, could be renewed and made more attractive to the residents.

Namely, the central districts could be restored and regenerated in a modern way while preserving their historical spirit. In this way the areas – presently undergoing degradation – would be utilized better and become more attractive, given the immediate accessibility of commercial and service outlets, transports and a unique natural quality. Another positive aspect of the proposition is that the compact character of the city would be enhanced, thus reducing the number and duration of journeys for the residents and increasing the rate of utilization of the city's public transport system.

3 Comparative Analysis of Two Concepts of the City's Spatial Development

The purpose of the analysis made in this section is to demonstrate that the spatial development concept of a 'diffuse city' prescribed for Bydgoszcz was not very effective for the residents or the municipality. For the purpose of the analysis comparisons were made of the number of possible journeys and their costs, durations and distances for the two possible spatial development scenarios:

- The 'diffuse city', which assumed the development of the city according to the Study of the Conditions and Directions of Spatial Development of Bydgoszcz, and
- The 'compact city', which assumed the development of the city according to the proposition of the authors of this paper.

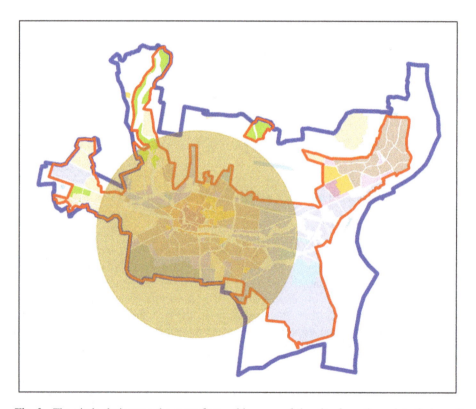

Fig. 2. The circle designates the most favourable areas of the city from the point of view of housing and socio-economic effects

The following assumptions were considered in the analysis:

1. The calculated costs of transport (verified for one selected year, 2015) were only applicable to the costs borne directly by the travellers and comprised the following:

(a) In the case of individual transport:
   (i) Operating costs of the vehicle,
   (ii) Travel time costs;
(b) In the case of public transport:
   (i) Ticket costs,
   (ii) Travel time costs.
2. All the unit costs were established according to the Blue Book – Road Infrastructure, July 2015 [15] (except for the costs of tickets which were based on actual prices).

Simulation analysis performed using the digital transport demand model revealed a significant increase in the cost of travel to be borne by the city residents if the spatial development scenario assumed in the approved plans was implemented (see Fig. 3). The average annual increase in the costs of travel with the assumed spatial development of the city in the years 2020–2050 exceeds 5.5 million EUR, as compared with the expected costs in the 'compact city' scenario, and takes approximately a third of the city budgetary resources earmarked for transport infrastructure. At the same time it should be noted that the extra costs rise in successive years of the forecast, from 2 million EUR in 2020 to nearly 9 million EUR in 2050, which is an equivalent of approx. 50% of the city funds available for investment in the local transport infrastructure. During the 30 years after 2020 the additional costs borne by residents only would exceed 172 million EUR, which is enough to fund three major investment projects. However, it must be stressed here that financial benefits of the 'compact city' concept are not yet considered. The benefits for the city budget would result from much smaller expenditure on transport investments (construction and extension of the road network, extra funds for public transportation infrastructure and facilities, extra funds for the rolling stock etc.) and lower costs of operation and maintenance of transport systems (e.g. public transport services).

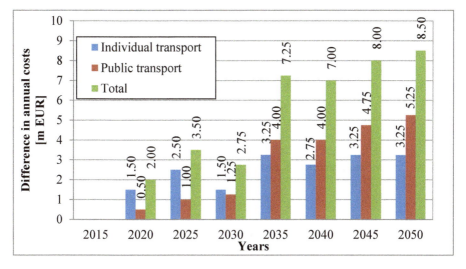

Fig. 3. Difference in the annual costs of travel by individual and public means of transport in two city development scenarios: the 'Diffuse City' and the 'Compact City' (forecast for 2020–2050)

The results of simulation analyses, presented in Table 1, indicate that the implementation of the 'diffuse city' development scenario – instead of the 'compact city' scenario – would bring about an enormous increase of the total travel distances in the municipal road network (from 28.5 m to 32.0 m vehicle-kilometers per year in the case of private transport, and from 4.6 m to 18.8 m passenger-kilometers per year in the case of public transport) and of the total travel time (from 520.2k to 639.3k vehicle-hours per year in the case of private transport, and from 124.2k to 667.2k passenger-hours per year in the case of public transport) in the years 2035–2045, according to the forecast. Moreover, the number of car journeys would increase by 0.8–1.2 m per year with a simultaneous decrease in the use of public transport by 0.5–1.2 m per year. In the case of public transport, the transfer rate would also rise by more than 4%. It should be noted that the most distant period of time of the forecast (i.e. the years 2035–2045) was purposefully selected for the presentation of the results, as that was the time horizon in which the city assumed the completion of the approved spatial development plans.

Table 1. The differences in the annual number, distance and time of trips in the city road network in the case of the "Diffuse City" scenario, as compared to the "Compact City" scenario

Year of the forecast	Private transport			Public transport			
	Total distance travelled [million vkm/year]	Total [thousand vh/year]	Total distance travelled [million trips/year]	Total distance travelled [million pkm/year]	Total time [thousand ph/year]	Total number of trips [million trips/year]	Transfer rate [%]
2035	28.5	639.3	1.1	4.6	124.2	−1.0	4.3
2040	33.5	562.2	1.2	6.4	105.9	−1.2	4.4
2045	32.0	520.2	0.8	13.8	667.2	−0.5	4.4

vkm – vehicle-kilometres,
vh – vehicle-hours,
pkm – passenger-kilometres,
ph – passenger-hours

4 Conclusion

Today it seems to be the perfect moment in the history of cities in Central Europe countries to implement – under the idea of sustainable development of the city and its transports – the model of a 'compact city' with a sensible spatial development policy which might provide for green spaces and open public spaces surrounded by urban developments. As demonstrated in Sect. 3, such a policy brings considerable economic and social advantages for the municipality and its residents.

What are the expectations of people seeking settlement in a medium-sized city in the era of globalisation and information society? They expect and demand literally everything, which – good career and education opportunities aside – includes the following:

- a very diverse choice of houses and apartments in terms of affordability and living quality,
- due consideration for their needs in terms of functional solutions offered in the residential buildings and their vicinity,

- short travel distance and time, and a reduced number and cost of necessary daily trips,
- availability of green spaces and safe public spaces in the neighbourhood (including safe traffic),
- a good choice of and easy access to commercial, service, healthcare, entertainment, leisure etc. facilities in the vicinity.

Therefore, the city housing policy should be integrated with the long-term transport development policy. The authors believe that it should meet the following requirements:

1. Creation of a compact city, proportionally developed, with low costs of living and good accessibility to all its areas, especially green and user friendly ones.
2. Promotion and stimulation of housing investments in the lands surrounding the city centres. To this end, old districts should be considered for conversion and regeneration and strategic reserves should be found.
3. Prevention of the urban sprawl towards its outskirts by making suitable provisions in the local land development plans (zoning plans) to restrict development of multi-family estates in those areas.
4. Development of financial and administrative instruments to influence investors' decisions on the location of their investment projects through a system of incentives (discounts, tax reliefs), assistance and support in administrative procedures and the promotion of particular city areas as the best for housing.
5. Investment in private and public transport systems in the areas of a high investment potential, to increase the quality of location and the value of land.
6. Zoning of the development density (including housing developments) in the local development plans to ensure the highest values of allowed density within the nearest vicinity of already existing transportation routes and public transport lines to use these transport potential.

The key objective of the housing policy described above should be to manage the spatial development of the city in the most beneficial direction. This requires a thorough multiple-criteria assessment of a number of development scenarios using state-of-the-art IT tools, such as 'transport demand models' used by the authors for this study, and the results of market research into the residents' preferences, that enable to understand residents transportation behaviour and their actual transportation demands.

It should be underlined that the analysis contained in this paper, as well as in the Transportation study for the city of Bydgoszcz, has already contributed to some advantageous corrections of the spatial development plans of the city in question. Decisions have been taken to change the land-use policy, and to implement above-mentioned recommendation for the present and future activity. The areas for housing places are recommended and promoted by the local city authorities. More over in the city centre a few attractive buildings with offices and accommodations have already been built, and the next one are planned.

References

1. European Commission: EU Transport in Figures. Statistical Pocketbook 2016. Luxembourg (2016)
2. U.S. Department of Transportation: Passenger Travel. Facts and Figures 2016. Bureau of Transportation Statistics, Washington (2016)
3. European Union: Cities of tomorrow. Challenges, vision, ways forward. European Commission, Directorate General for Regional Policy (2011)
4. Banister, D.: Unsustainable Transport: City Transport in the New Century. Taylor & Francis, Hoboken (2005)
5. Marshall, S., Banister, D.: Land Use and Transport: European Research Towards Integrated Policies. Elsevier, Amsterdam (2007)
6. Central Statistical Office of Poland, Statistics in Transition (SiT) in 2015. Warsaw, Poland (2016)
7. Graham, S.: Constructing premium network spaces: reflections on infrastructure networks and contemporary urban development. Int. J. Urban Reg. Res. **24**(1), 183–200 (2000)
8. Macioszek, E., Lach, D.: Analysis of the results of general traffic measurements in the west pomeranian voivodeship from 2005 to 2015. Sci. J. Sil. Univ. Technol. Ser. Transp. **97**, 93–104 (2017)
9. Macioszek, E., Lach, D.: Comparative analysis of the results of general traffic measurements for the Silesian Voivodeship and Poland. Sci. J. Sil. Univ. Technol. Ser. Transp. **100**, 105–113 (2018)
10. Banister, D., Berechman, J.: Transport Investment and Economic Development. UCL Press, London (2003)
11. Macioszek, E., Lach, D.: Analysis of traffic conditions at the Brzezinska and Nowochrzanowska intersection in Myslowice (Silesian Province, Poland). Sci. J. Sil. Univ. Technol. Ser. Transp. **98**, 81–88 (2018)
12. D'Argent, N., Beringer, J., Tapper, N.: Planning for the compact city; an assessment of Melbourne@ 5 million. In: 7th International Conference on Water Sensitive Urban Design, Melbourne Cricket Ground, 21–23 February 2012, pp. 781–785 (2012)
13. Barrella E.: Transportation Planning for Sustainability Guidebook, Georgia Institute of Technology, pp. 31–39 (2011)
14. PTV VISUM 16 Fundamentals, PTV AG, 724–734, Karlsruhe, Germany
15. Jaspers, Blue Book Road Infrastructure, pp. 47–66 (2015)

The Impact of Cities' Spatial Planning on the Development of a Sustainable Urban Transport

Grzegorz Dydkowski[✉]

Department of Transport, Faculty of Economics,
University of Economics in Katowice, Metropolitan Transport Authority (ZTM),
Katowice, Poland
grzegorz.dydkowski@ue.katowice.pl

Abstract. Sustainable development of cities, including also sustainable development of urban transport, starts already at the stage of city planning and spatial planning, adoption of specific assumptions in the field of spatial development intensity and functional-spatial structures of the city. At the moment decisions are made on the amount of utilised resources, such as the land, on the expenditure on the city transport infrastructure, or the size of transport needs and the consumption of energy related to transport. Intensive spatial development of urban areas in general reduces the demand for individual transport making them on foot, by bicycle, or by public transport. On the other hand suburbanisation processes existing universally result in increased demand for the road infrastructure and force the necessity to use personal cars. This paper made an attempt to present the relationship between social-economic and spatial structure transformations, the population density and intensity of cities development, and the use of infrastructure and the way of providing by the urban transport. In particular it is necessary to pursue implementation of solutions increasing the utilisation of the existing infrastructure and relationships between the mentioned phenomena occurring in cities and in suburban areas and the division of transport tasks and the effectiveness of the public transport.

Keywords: Urban transport system · Spatial management of cities · Sustainable urban development

1 Introduction

The issues of city sustainable development are now one of the basic topics, considered by numerous papers related to city functioning and management, spatial and city planning, environmental protection of cities, ecology, urban transport operations, and ensuring mobility in cities. Basic decisions related to the spatial planning apply to functions and purpose of individual city areas and to the location of technical infrastructure. Such decisions affect arrangement of traffic generators in the city space and solutions in the city transport system, including the transport tasks division. The decisions made on the city spatial planning have a long-term influence on the living conditions in cities, on functionality, attractiveness, costs of city infrastructure operation, transport needs availability and volume.

In big cities possibilities of transport infrastructure development, especially roads for cars, become increasingly limited, causing that the transport factor becomes the basic one, conditioning possibilities of their development. As a result the approach to the transport planning is no longer based on as wide and fast as possible development of the network, the maximum balancing of the demand and supply became the priority. Because of that the attitude to space has changed, from the place to be occupied by the developing transport networks to the place, which adapts to the actual needs of transport users, taking into account local demand management and integrated network planning. To a greater and greater degree the undertaken projects focus on the demand management, and to increasingly smaller - on a new infrastructure supply. In the process of city social-economic and spatial structure transformations also the process of suburbanisation is observed with various intensities, as the effect of residents, businesses, commercial and service centres migration from central zones into suburbs.

The issues of the urban transport sustainable operation in the context of city spatial planning and functional-spatial structures were already addressed. In Poland the research and papers by e.g. J. Podoski, S. Dziadek, K. Roszko, J. Malasek, W. Suchorzewski, A. Rudnicki, J. Bogusławski, M. Rościszewski, and J. Wesołowski may be mentioned. Among the research carried out abroad in particular papers by P. Newman and J. Kenworthy [1, 2], R. E. Brindle, P. Naess, Pushkarev, J. Zupan, T. Haasa, D. Farr, or M. Mostafavi may be specified. This issue is also discussed in documents defining policies and strategies of city development as well as of transport and mobility in cities, and also documents adopted on the national [3] and European Union [4] scale.

This paper is an attempt to make a kind of synthesis related to the relationship between social-economic and spatial structure transformations, the population density and intensity of cities development, and the use of infrastructure and the way of providing the urban transport service. In the study basic methods of economic research were used, in particular the method of critical literature review, which provides, among others an assessment of the state of scientific knowledge in the field of transport and spatial development. Methods of analysis and synthesis were used in the study as well. These methods were used to understand the impact of spatial development on transport system and to evaluate of existing solutions as well as to formulate final conclusions. In particular it is important to look for solutions increasing the usage of the existing infrastructure and aimed at sustainable and greater use of environment-friendly forms of movement, such as walking, cycling, and moving by public transport.

2 Relationships Between Spatial Planning and Urban Transport Systems Planning

There are numerous concepts of cities development, also processes of their transformations and development over years proceed in various ways. The city size, its social-economic and functional-spatial structure changes, it goes through social and demographic processes, also expectations related to the living conditions in the city change. From the transport point of view the intensification of spatial planning is the basic feature of the urban space. The concentration enables to intensify contacts as well as the social, economic, and cultural exchange and by that - the origination of the added value and civilisation development. Economic benefits of greater concentration consist in increased productivity, functionality, and attractiveness, related to lower infrastructure costs and reduced transport needs [5]. The 20th century experience has shown that the population density and intensity of development support productivity, functionality, and the value of city space. The concentration of development brings obvious economic benefits, such as lower infrastructure costs, lower transport costs, and economic justification of the urban transport development as well as time savings at movements [5].

Sustainable mobility requires starting actions, which should result in reduction of the need to travel (reduction of the number of journeys), limitation of covered distances, change of the means of transport into a more ecological one, ensuring a higher transport efficiency, and improvement in availability. Such objectives can be accomplished by the application of transport policy and spatial planning instruments, by substitution of physical transport by other forms of activity, and by implementation of technological innovations contributing to growing effectiveness of transport systems [6]. Figure 1 presents the location of the transport development planning (including demand management, transport planning, and traffic management) with respect to the spatial planning and city development planning as well as regions master planning and the concept of the country master plan, and also a set of tools, which should influence transport users to adopt more ecological attitudes.

The traffic management is aimed at the optimisation of roads capacity via the traffic streams control. In the demand management usually 'soft' measures to affect transport users are applied as a rule, and their application should encourage more ecological behaviour. 'Hard' measures are understood as the infrastructure and the regulatory side of transport planning (legislation, regulations, taxes and tariffs), which usually are compulsory for the user [7].

In the processes of city spatial planning and of transport systems planning it is necessary to ensure solutions, which reduce the traffic volume, reduce the congestion,

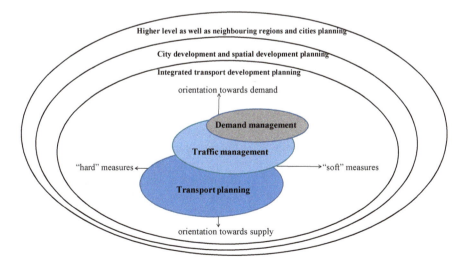

Fig. 1. Transport development and demand management planning framework (source: [6, 8])

and which will make the transport in the urban and suburban areas more productive, safer, and more environment friendly. Because of the way of proceeding, two systems of spatial planning and transport planning integration strategy can be distinguished [9]:

- land use planning policies aimed at reduction of needs to travel - these are policies aimed at formation of new centres or regeneration of the developed city areas, transformation of their tissue and reduction of the urban sprawl (housing estates, places of work, etc.);
- transport policies aimed at improving access through a wide choice of transport alternatives - such policies assume changes in the transport system to improve access via alternative methods of travelling (public transport, pedestrian traffic, bicycle traffic, flexible transport services, car sharing, etc.) and via stimulation of revitalisation of densely populated areas and of those of mixed use within the city.

Transport systems planning, affecting the residents transport behaviour, and decisions from the field of city land use must be mutually coordinated. The development of city areas is strongly related to transport. The basis of traffic analyses and forecasts consist of a rule, that the existence of distances between places of human activity creates the need of travelling and of goods transport - so the development of suburban areas is related to a greater mobility of residents. The following are affected by the spatial planning:

- functional-spatial structure of the city, mutual arrangement of city functions,
- land use intensity, which next conditions the passenger flows concentration and, as a result, types of means of transport used in the urban transport,
- spatial arrangement of means of public transport routes, integration nodes, transfer points,
- streets and roads capacity and car parks locations.

The arrangement of places of residence and work, the population density, and the diversification of land use intensity create premises for serving by means of transport of varying capacities. The consideration of transport services during the spatial planning, creation of new centres along public transport - in particular rail transport - axes, and development within a walk to a public transport stop allow to ensure convenient access - with possible transfers at integration nodes. Solutions facilitating multimodality and the use of various means and methods of transport in such situations become an instrument to transform transport systems and to adapt them to changes proceeding in the land use of individual areas.

The occurring social and demographic processes - ageing of people and increasing number of small and one-person households - contribute among other things to growing demand for new apartments, to pressure on the development of suburban areas (so-called suburbanisation), as well as to expectation of improved living conditions in central areas of cities. Cities transform to improve living conditions for their residents - better housing conditions, larger apartment areas, ensuring green spaces, places of recreation as well as easy access to goods and services. On the one hand it causes increasing demand for the land and space in cities and the interest in suburban areas, but on the other hand a good access to places of work, education, recreation, services, and other activities is expected to be guaranteed.

The process of 'city sprawl' progresses in the majority of urbanised areas. The highest pace of population growth occurs in satellite cities and in suburban areas, so far with a low density of population. The suburbanisation is commonly treated as a stage of urbanisation process, however, it is difficult to determine, what is characteristic of for just this phase, because it is perceived as a multidimensional and complex phenomenon, and it is impossible to indicate features clearly distinguishing this process. This is a part of general process of developing city systems complication, both in a functional (urban functions mixing, polycentricity), and also in social-cultural and economic sense [10]. Modern urbanisation processes, comprising vast areas beyond city reach, frequently feature parallel deconcentration and re-concentration of population as well as economic and social activity, leading to the origination of multi-centric metropolitan regions. This results inter alia in growing domination of the transport and logistic system in the spatial and social-economic structure, combined with incommensurate growth of infrastructure (motorways, roads, ring roads, service stations, airport and its facilities, internal transport systems), which nodes gradually take away the functions from the old centre [10].

Suburbanisation processes substantially decrease the effectiveness of transport in European cities. This is related to growing dependence on car usage in those areas. Processes of suburbanisation change monocentric city areas in complex polycentric conurbations. Increasingly greater spatial-functional segregation is an important consequence of this process, resulting in making apartments, places of work, of trade, and of other services more distant. The suburbanisation and concentration of economic and service activities (e.g. shopping malls), which enable obtaining economies of scale, lead to a growing average length of travel and to travelling from one suburb to another. In general suburbanisation consists in increased travel distances and thereby in being more dependent on a car, because areas of low buildings, frequently also situated very chaotically and in a large area, are very difficult and inefficient to serve by the public transport as compared to individual journeys by cars.

However, also symptoms of re-urbanisation appear. An active policy of many urban areas renewal and revitalisation obtains some success in the form of withdrawing depopulation and stopping city centres destruction [9]. This may be assessed positively, especially from the point of view of transport systems efficiency. These are the city centres, where historically the public transport infrastructure is located, in particular of the rail and high-capacity transport, which can be utilised.

3 Location Decisions Related to the Land Use

The location decisions and the intensity of land use should be considered form the point of view of shortening the time and distance of travel, that is reduction of the transport intensiveness. This can be obtained through the transformation of town planning structures from mono-functional to multi-functional, pursuit of local balancing of places of residence, work, and service offer, preventing processes of settlement deconcentration into areas and reserving in master plans the areas for car parks and integration nodes.

Table 1 presents the impact of land use rules on the transport behaviour of residents. A high population density has a small impact on reduction of an average travel distance, at no increase in the travel cost; at the same time there is an assessment that a high employment density is strongly correlated with an average travel distance.

Attractive places in local surroundings have 'attracting' importance, hence they result in reduction of the travel distance. Outskirts are usually related to longer

Table 1. Influence of land use related decisions on the transport system.

Direction	Factor	Influence on	Expected results
Land use ↓ Transport	Population density	Travel distance	A higher population density itself will not result in shorter travel times. Places of work with places of residence mixing leads to shorter travel distances, if their costs grow
		Travel frequency	Small effects should be expected. At shorter journeys their number can grow
		Means of transport choice	Population density higher than some minimum is favourable for the public transport effectiveness. Walking and cycling will become more frequent only, if average travel distances are shortened (see above)
	Employment density	Travel distance	Concentration of places of work in a few centres usually results in increased average travel distances. The balance in the number of places of work and of residence in a given area results in shortening of the travel time only, when their cost is high
		Travel frequency	Small effects should be expected. At shorter journeys their number can grow
		Means of transport choice	Concentration of places of work in a few centres can result in reduced car usage, if effectively supported by the public transport. Popularity of walking and cycling will increase only, if they are short (see above)
	Density of neighbourhood	Travel distance	Attractive public spaces and diverse shops and service points can encourage to local journeys
		Travel frequency	At shorter journeys their number can grow
		Means of transport choice	Street layout - spaces for pedestrians and bicycle tracks - can increase popularity of walking and cycling
	Location	Travel distance	Longer travel distances are a feature of places in suburban locations
		Travel frequency	No effects should be expected
		Means of transport choice	Places close to public transport nodes usually show a higher share of public transport in travelling
	City size	Travel distance	Travel distance is usually negatively correlated with the city size
		Travel frequency	No effects should be expected
		Means of transport choice	Bigger cities can afford maintaining more efficient public transport systems, so the public transport should be more popular there

Source: [11]

journeys, so the travel distance can be determined as having a negative correlation with the city size.

Table 2 presents the impact of transport policy on decisions in the field of various functions and activities location in the city and on the transport behaviour of residents.

Transport decides about the land accessibility and better access increases the location attractiveness for all types of the land use, and thereby affects the city

Table 2. The impact of decisions related to the transport infrastructure on decisions related to the location and to the public transport usage.

Direction	Factor	Influence on	Expected results
Transport ↓ Land use	Access	Apartments location	Sites of better access to places of work, shops, schools, and places of recreation are more attractive for the housing development; they feature higher land prices and faster development. Local access improvement changes location decisions related to new residential buildings; access improvement in the entire city area results in more dispersed apartments location
		Industry location	Locations with better access to motorways and freight railway stations are more attractive for the industry development; this development is also faster in such places. Local access improvement results in a change of the industry development direction
		Office locations	Locations with better access to airports, to a fast rail station, and to motorways are more attractive for offices development; also land prices in such places go up. Local access improvement results in a change of the office sector development direction
		Trade location	Locations of better access for the customer and competition are more attractive to trade development, they have higher land prices and develop faster. Local access improvement results in a change of the trade development direction
Transport ↓ Transport	Access	Travel distance	Locations with better access to many places are generators of longer journeys
		Travel frequency	Locations with better access to many places are generators of bigger number of journeys
		Means of transport choice	Locations with good access for cars generate more journeys by this means of transport; locations with better access by public transport - generate more journeys by the public transport
	Travel cost	Travel distance	Increase in the travel cost results in reduced travel distances
		Travel frequency	Increasing travel cost results in a reduced number of journeys
		Means of transport choice	There is a strong relationship between the travel cost and the means of transport choice
	Travel time	Travel distance	Increase in the travel time results in reduced travel distances
		Travel frequency	Increase in the travel time results in a reduced number of journeys
		Means of transport choice	Means of transport choice heavily depends on the travel time

Source: [11]

development. It is necessary to draw attention to the fact that if accessibility in the whole city grows, then the settlement structure will be dispersed [11]. So the transport service concentration on selected routes is a factor allowing to shape the expected location decisions in the city. There is also a relationship between the quality of specific area transport service and the residents mobility. Good access increases the residents mobility, while the increase in the transport cost, extension of travel distance and time result in the reduction of the number of journeys. Also the travel time is a significant element, it affects the means of transport choice. It is visible that decisions on the method of moving result from many factors, they include the time of access to the stop, the travel time, cost, and other elements deciding about the comfort. Assuming that a part of journeys will be carried out in systems with transfers, it is necessary to not increase their cost and travel time as compared with direct journeys.

4 Intensification of the Existing Transport Infrastructure Utilization

For many years there were attempts to resolve – with varying success – the problems resulting from the increasing number of vehicles and growing traffic on roads and streets primarily through the construction of new road sections. The issues related to the traffic management and utilization on roads were actually reduced to road marking and to controlling the traffic so that movements would be safe. The number of vehicles was systematically growing, the amounts of roads – slower, also the traffic volume on the roads was increasing [12]. The construction of new roads did not reduce the roads congestion or reduced it only for a very short time – a few months or years. After such period the situation was returning to the starting point. However, it is necessary to emphasise, that the maximum utilization of roads, streets or car parks occurs only in a part of day, at selected days of week, or during a few months per year. The unevenness of roads utilization depending on the direction of driving should be added to that – for example in the morning hours roads leading to city centres are much more loaded than in opposite directions, then during the afternoon hours the vehicles traffic concentrates on the opposite directions – on driving to residential as well as trade and service areas. Also the car parks occupancy is similar – car parks close to places of work are filled during the working hours, while car parks at housing estates during the evening and night hours, and at shopping and service centres and at places of recreation – during afternoon hours and weekends. Therefore now, apart from safety elements, attention is more and more often drawn to the issues of infrastructure utilization, using at that the utilization technologies (telecommunication and IT). By undertaking actions from the field of road traffic management it is possible to use the infrastructure in a better way, and to ensure smoother movements of vehicles and pedestrians on roads and streets. Apart from benefits related to the shortening of the travel time and the reduction of the fuel consumption it allows also to reduce the emission of harmful compounds formed as a result of combustion engines operation and of noise, because they are related primarily to vehicles' starting.

The traffic volume on individual road sections an hence the roads' load features a high variability over time. As a result the application of only static solutions, i.e. such, in which on the current basis there is no response to a changing situation on the roads, is insufficient. Hence cities plan or implement to a various extent traffic management systems utilization the IT technologies. In general, the traffic volume and situations on the most important intersections or other points are monitored continuously and the data is used for the area control of traffic lights on intersections and of variable message signs. Moreover, e.g. on internet portals, on the radio, or displaying on various signs and displays, the information about the traffic conditions on the roads is provided in real time, as well as on the existing difficulties in traffic, temporary traffic lane closures, so that road users could plan in advance their journey and in case of obtaining information about difficulties – to choose another route, Traffic management systems can also facilitate driving for vehicles privileged in the traffic or for the urban transport vehicles. The comprising of the means of urban public transport by the traffic management systems allows to obtain many benefits and thereby to improve its competitiveness against travelling by car. In particular the benefits include:

- improved punctuality and regularity of means of public transport journeys,
- reduced travel time (shortening the time of journeys and times of waiting for a vehicle, as well as the time planned for transfers),
- as a result of improved regularity also a more even occupancy of vehicles is obtained, the number of situations, where vehicles are overfilled, is reduced,
- passengers are provided information about disturbance in traffic, if any, and about a possibility of journey continuation.

In general, the traffic management systems improve the quality of urban public transport services, the confidence in such method of travelling as well as, among other things by the improved regularity of journeys and shortened travel times, improve the utilization of urban public transport vehicles.

Personal cars only during a small part of a day or week are used to travel – the remaining time consists in parking. Traffic management systems in cities can also provide information about availability of parking places, this is primarily aimed at reducing the volume of traffic related to looking for a parking place, at decreasing the number of vehicles entering central areas of cities (the entry may be pointless, if it is not possible to park the car), and at better utilization of car parks operating under Park & Ride systems. A part of traffic management systems should also comprise systems allowing, based on developed traffic models, to prepare long-term forecasts of traffic on the road network or short-term simulations of road network load in various situations, e.g. of mass events, closure of selected road sections, closure of traffic lanes, increased share of people moving by the public transport. They allow to evaluate the impact of the mentioned situations on the increased volumes and travel times, and to make the optimum choice of proceedings option so as to reduce the nuisance to drivers and to the whole city.

5 Principles of Transport Needs Serving Depending on the City Size

Urbanised areas of low spatial development intensity, are served to a prevailing degree by personal cars (Table 3). Low spatial development intensity means at the same time higher energy consumption in transport per capita. With increasing development intensity the urban public transport is used to travel to a bigger extent. This becomes especially important in the case of big and densely developed cities. Then also the role of transport integration grows, as a factor allowing to reduce the share of cars in favour of the public transport, in particular high-capacity means of transport.

Table 3. Typology of average spatial development intensity in a metropolitan area vs. selected transport parameters.

Specification	General net spatial development intensity in the city		
	Small <25 m + p/ha	Average 50–100 m + p/ha	High >250 m + p/ha
Total modal split			
Car usage (km/person/year)	>10,000	No data	<5,000
Public transport usage (journeys/person/year)	<50	No data	>250
Fuel/petrol consumption in transport (MJ/person/year)	>55,000	35,000–20,000	<15,000
Representative situations	Metropolises in Northern America and Australia	European Metropolises	Asian metropolises and the biggest urban centres worldwide

Spatial development intensity (m + p/ha): number of residents (m) and of jobs (p) per 1 hectare (ha) of land in the city (excl. green areas, water reservoirs)
Source: [1]

The relationship between the spatial development intensity and the division of the transport work and the energy consumption and the city environmental burden is presented in Fig. 2 [13]. The integration of the urban public transport requires a diversified approach depending on the city size. In the smallest settlements the local transport needs are satisfied by entities rendering transport services in inter-municipal relations, which are pretty often classified in the subject literature as regional transport. With increasing city size the service by an urban transport system becomes universal.

In small cities, the urban public transport services are provided most frequently by one entity, and city buses are the means of transport. Such entity often operates also lines going beyond the city borders, to production plants situated there, shopping centres, or ensuring access to places of residence or to neighbouring cities.

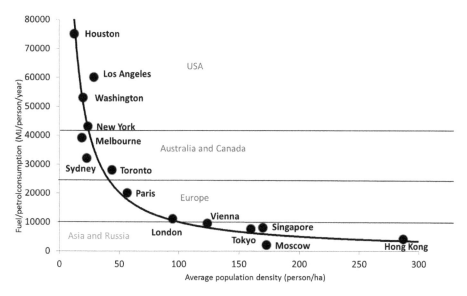

Fig. 2. The relationship between the density of buildings and the consumption of petrol (source: [5, 14])

There is one public transport operator (frequently combining the urban, suburban, and regional transport), one traction, a high percentage of journeys are carried out without transfers, which results from a small number of traffic origins and destinations and from a possibility to ensure direct connections by the bus transport. There is a high share of movements by cars, enabling direct journeys from the origin to the destination. The transport integration applies primarily to convenient transfers between the urban and regional public transport systems, and between the public transport and movements by individual cars. Because of a high share of single-track vehicles at travelling to stations, places guaranteeing their safe leaving at the railway or bus stations should be ensured.

With increasing size of the city the volumes of passenger flows on individual route sections become diversified, hence various means of transport are used for the service, frequently managed by various operators. There are separate management and financing systems for the urban and regional public transport. Because in such cities the public transport system consists of subsystems (city railway, trams, buses) overlapping and supplementing each other. Moreover, a higher transfer rate results from the multitude of traffic origins and destinations existing in big cities, from the necessity to concentrate passenger streams on the main transport axes and to restrict movements by cars - partly by combining movements by car with journeys by the public transport. In big cities the transport integration, apart from organisational and tariff solutions, consists in the spatial integration of various transport branches and means of transport. The integration of transport, starting from decisions at the stage of town planning, becomes a tool allowing to:

- concentrate passenger streams on efficient transport branches (rail systems),
- use cars and bicycles in areas of small population density, where an effective operation of public transport systems is also difficult,
- use a car, hence starting a journey in the immediate vicinity of the place of residence (without walking to a stop) in the case of Park & Ride journeys and convenient transfer centres - reduction of walks,
- combined with information, to optimise the movement route due to the shortest time or meeting the assumed travel time (possible transfers at various places),
- achieve benefits for the environment, related to the use of means of rail transport and to match better the supply of services.

In the process of master planning the areas (spaces, locations) are determined, intended for transport functions, as well as the transport system of the city, that is the shape of roads, routes, and public transport lines network together with integration nodes, stations, termini, stops, and more important car parks. At this stage, from the transport integration point of view, it is important to determine integration (transfer) nodes, both those intended for transfers between means of public transport, and between cars and the public transport. The areas of transfer centres and the applied solutions should facilitate transfers performance.

6 Summary

Sustainable development of cities, including also sustainable development of urban transport, starts already at the stage of city planning and spatial planning, adoption of specific assumptions in the field of spatial development intensity and functional-spatial structures of the city. At this moment decisions are made on the use of resources, such as the land, expenditure on the city transport infrastructure, or the size of transport needs and the consumption of energy related to transport. Compact cities, with ordered functional-spatial structure, developing along pro-ecological and efficient transport systems, create conditions for the public transport operation and reduction of personal cars traffic. An intense development reduces the transport needs, creates possibilities for convenient movements on foot or by bicycles and to use the public transport. It generates also smaller demand for energy and thereby a smaller emission.

The pursuit to possess an own house among greenery, easier access and lower prices of land in the suburban areas, the pursuit to achieve economies of scale and locating there residential buildings and frequently also big commercial and service centres, results in the transfer of population and of certain activities into suburban areas. Such areas are difficult to be efficiently served by the public transport, they generate a significant demand for the infrastructure and the car traffic. It is difficult to prevent such a phenomenon - it became possible due to availability of personal cars - it existed and exists in many cities worldwide, and actually it should be captured within some framework, so that it is not chaotic, and to reduce its negative effects, and also over time integrate the suburban areas and merge them into one consistent city organism.

IT technologies play a significant role in increasing the capacity of the existing transport infrastructure and the urban transport systems, as well as in the environmental protection. The implementation projects in the field of road traffic management provide a possibility to use the infrastructure in a better way, and to ensure smoother movements of vehicles and pedestrians on roads and streets. This allows obtaining benefits in the form of shorter travel times and the reduction of fuel consumption, and that means a reduction of the emission of harmful compounds formed as a result of combustion engines operation and noise, because they are related primarily to vehicles' starting.

References

1. Newman, P., Kenworthy, J.: Sustainability and Cities. Overcoming Automobile Dependence. Island Press, Washington D.C. (1999)
2. Newman, P., Kenworthy, J.: Urban design to reduce automobile dependence. Opolis 2(1), 35–52 (2006)
3. Koncepcja Przestrzennego Zagospodarowania Kraju 2030, uchwała nr 239 Rady Ministrów z dnia 13 grudnia 2011 w sprawie przyjęcia Koncepcji Przestrzennego Zagospodarowania Kraju 2030, Monitor Polski z dnia 27 kwietnia 2012, poz. 252. (in Polish)
4. The Green Paper Towards a new culture for urban mobility, Commission of the European Communities, Brussels, on 25.0.2007, COM(2007) 551 final; Together towards competitive and resource-efficient urban mobility, Communication From The Commission To The European Parliament, The Council, The European Economic And Social Committee And The Committee Of The Regions, Brussels, 17 December 2013 COM(2013) 913 final
5. Stangel, M.: Kształtowanie współczesnych obszarów miejskich w kontekście zrównoważonego rozwoju. Wydawnictwo Politechniki Śląskiej, Gliwice (2013). (in Polish)
6. Koźlak, A.: Kierunki zmian w planowaniu rozwoju transportu w miastach jako efekt dążenia do zrównoważonego rozwoju. Transport Miejski i Regionalny 7(8), 39–43 (2009). (in Polish)
7. Pressl, R., Reiter, K.: Transport behaviour management, Research for Sustainable Mobility, European Commission 2003. http://www.eu-portal.net
8. Ahrens, G., Shone, M.: Integrated transport planning leads to mobility management visions for cross-border co-operation. In: ECOMN 2006 Conference Paper, Groming (2006)
9. Achieving Sustainable Transport and Land Use with Integrated Policies, Final Report, Transplus, European Commission (2003)
10. Heffner, K.: Proces suburbanizacji a polityka miejska w Polsce. http://dx.doi.org/10.18778/8088-005-4.05. (in Polish)
11. Transport a zagospodarowanie przestrzenne. Wyniki finansowanych przez UE prac badawczych w dziedzinie transportu miejskiego (2003). http://eu-portal.net. (in Polish)
12. Internet website of the General Directorate of National Roads and Motorways. http://www.gddkia.gov.pl/pl/987/gpr-2010
13. Bovy, P.H.: Integracja planowania urbanistycznego i transportu a bardziej zrównoważony rozwój ruchliwości. Biuletyn Komunikacji Miejskiej 58 (2001). (in Polish)
14. Newman, P., Kenworthy, J.: Cities and Automobile Dependence: An International Sourcebook. Gower Publishing (1989)

The Role of Intelligent Transport Systems in the Development of the Idea of Smart City

Wojciech Lewicki[1], Bogusław Stankiewicz[2], and Aleksandra A. Olejarz-Wahba[3(✉)]

[1] The Faculty of Economics, West Pomeranian University of Technology, Szczecin, Poland
Wojciech.Lewicki@zut.edu.pl
[2] The Faculty of Economics, The Jacob of Paradies University, Gorzów, Poland
bstankiewicz@zut.edu.pl
[3] The Faculty of Economics, The University of Warmia and Mazury in Olsztyn, Olsztyn, Poland
wahba@wp.pl

Abstract. As the available literature indicates, Smart city is an idea in which the city uses, among others, information and communication technologies to increase the interactivity and efficiency of transport infrastructure and the transport supra-structure. Although specialists dealing with the issues of smart cities use different definitions, they agree in one thing, the availability to services offered by intelligent transport systems is the key to development of this concept. On this basis, the authors have attempted to characterize the currently available services in the field of intelligent mobility in relation to road transport. The aim of the article is to show that the city, in order to be perceived as "intelligent", should first of all try to effectively use the solutions that are currently available on the market in the field of intelligent transport systems.

Keywords: Smart city · Services · Intelligent mobility · Intelligent transport systems · Road transport

1 Introduction

As available literature on the subject indicates, Smart City is an idea in which investments in human and social capital, supported by the development of modern infrastructure, will be the basis for sustainable economic development and the building component of the so-called high quality of life for citizens living in urban areas [1–3]. However, as postulated by many authors, the implementation of such an ambitious plan will require not only significant financial outlays, but also the application of appropriate technological solutions, including those in the field of intelligent mobility [4, 5].

It has been stressed repeatedly that the discrepancy in the awareness of vehicle users is clearly visible in the process of developing intelligent mobility in relation to road transport [1]. Many authors put forward the thesis that most of the cases do not concern expectations in relation to the infrastructure, or rather the means of transport itself, that is, their equipment [6]. As available reports and studies indicate, a significant part of users do not use the currently available services in the field of intelligent mobility because they are not aware that their vehicles are equipped with individual devices and telematic applications [7]. If intelligent mobility is to support the smart city idea, it is obvious that the first step should be to convince vehicle users to use solutions that are already offered by their vehicles.

Therefore, according to the authors, one of the most current challenges in the implementation of the smart city concept is to identify as fast as possible the benefits of both economic, social and ecological nature, resulting from the application of these solutions in practice. Such educational activities may lead to the increased interest in the already available solutions in the field of intelligent mobility.

As indicated by the analysis of the available literature, despite the fact that the smart city idea is not an innovative concept, there are is no literature that would describe the problems of available services related to intelligent mobility in relation to road transport in an interdisciplinary approach. What makes this subject even more interesting and worth considering in order to diagnose the positive effects and rightness of the actions taken in this matter.

Therefore, the discussion of these issues seems to be the proper identification of the research problem, covering both the aspects of intelligent transport systems and the idea of Smart City.

The approach presented above became for the authors the basis for adopting boundary conditions and methods of conduct aimed at attempting to dimension the role of intelligent transport systems in the development of the idea of Smart City by:

- Bringing the idea of Smart City and the role of intelligent mobility closer.
- Characteristics of the available services in the field of intelligent mobility in relation to road transport.

The fundamental goal of the article is to show that the city, in order to be perceived as "intelligent", should first of all try to effectively use the solutions currently available on the market in the field of intelligent transport systems. The implementation of this task required the authors to apply appropriate research methods and techniques, such as the method of induction and deduction, as well as analysis of available literature sources and observations of market reality. The direct recipients of the results of the research, apart from researchers dealing with the issues of smart city and intelligent transport systems, will also be entities that carry out transport activities and vehicle users themselves.

2 The Smart City Idea and Intelligent Mobility

In Nowadays, urbanization processes are under the influence of globalization and technical and technological progress. Cities, in order to fulfil their function as an attractive place to work and live, should correspond to social needs and business requirements [8]. Therefore, the challenge for city authorities and policy makers at the national and global levels is to take action and find solutions to the question of how cities can become intelligent in the environment of changing conditions and many unknowns? The answer is to make cities "more intelligent" through effective management of resources and infrastructure, increase environmental performance and appropriate legal and administrative decisions, which will improve the quality of life of citizens. This state can be achieved by effective use of the Information and Communication Technologies (ICT), which are able to provide solutions for cities that are citizen- and environment-friendly and economically viable [7].

As indicated by observations of market reality, several dozen years ago, cities began the process of implementing selected solution in the area of ICT in order to provide citizens with better services that improve their quality of life. The evolution of cities takes place from centres functioning traditionally, through the subsequent stages of digitization, communication and information, up to the concept of the so-called smart cities. Currently, there is a discussion in the world of science about how to define smart cities. Despite the existing consensus regarding the definition of a Smart City, which includes innovations in city management, services or infrastructure, the general, unambiguous and extensive definition of a Smart City has not yet been developed. There is a wide range of definitions answering the question of what a Smart City really is [9]. Initially, it can be assumed that the concept of a Smart City refers to a certain idea or style of the city's functioning. The word smart can be translated into Polish as "intelligent", "clever", "ingenious", "apt", "resourceful". In the context of the city, none of these words in Polish sounds clear, and thus does not reflect the actual idea of a Smart City, defined, e.g., in the documents of the British government as a kind of civilization process. However, because numerous new "smart" technologies tend to be translated into Polish as intelligent, this term is also included in this article. Thus, depending on the perspective of the research area, several definitions of a smart city can be distinguished. On the one hand, there is a set of definitions focusing mainly in the urban aspect (technology, environment), omitting other factors related to the city. This group of monothematic descriptions may cause a misunderstanding of the main goal of a Smart City, which is to provide a new approach to city management, based on the combination and consideration of all factors and conditions affecting life in the city. Active cooperation of citizens with the city administration increases the possibility of solving key social and urban problems. The improvement of only one part of the urban ecosystem does not solve all problems and challenges related to city management [10]. On the other hand, some authors emphasize the importance of combining all urban aspects in the context of building a Smart City. The combination of social, infrastructure, institutional challenges and the search for solutions at the same time is reflected in the concept of Smart City [11]. Infrastructure and technology are the central elements of a smart city, but it is the communication, cooperation and integration of all elements and systems that makes cities smart. Based on the examples of Smart City definitions quoted above, it can be concluded that this concept assumes a comprehensive approach to the city in terms of management, residents and development. The definitions mentioned above essentially emphasize the balance of technological, economic and social factors related to the urban ecosystem. They reflect a holistic approach to the city's challenges related to the use of new technologies that cause the change if interactions between processes and participants of the city. On the basis of the examples of Smart City definitions quoted above, it can be concluded that this concept assumes a comprehensive approach to the city in terms of management, residents and development. Currently known definitions emphasize the balance of technological, economic and social factors related to the urban ecosystem. They reflect the holistic approach to the city's challenges related to the use of new technologies that change the interaction between the processes and participants of the city. Based on the above-presented considerations, it can be concluded that the use of the Smart City model can lead to better planning and city management, contributing to ensuring lasting sustainable development.

As indicated by the literature on the subject, research works analysing the urban environment focus on six main areas of the city [4, 8, 12]. Intelligent mobility is one of these areas. Many authors identify this concept with intelligent transport systems, the idea of which is to combine advanced information and communication technologies with the existing transport infrastructure to improve security, protect the natural environment and increase the efficiency of transport processes (Fig. 1).

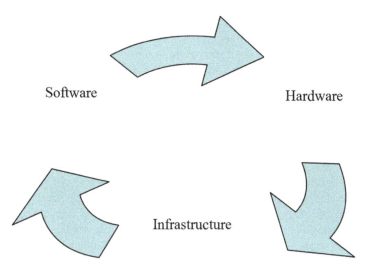

Fig. 1. Components of intelligent mobility

At this stage of considerations, it is worth mentioning that the available research shows that the use of ITS in road transport leads to:

- increasing the capacity of transport infrastructure (by an average of 22.5%).
- improvement of road traffic safety (a significant reduction of accidents by 50% in the built-up area).
- significant reduction of travel times and energy consumption (by about 60%).
- improvement of the quality of the natural environment (reduction of exhaust emissions by an average of 40%),
- improving the comfort of travelling and the conditions of traffic for drivers and people travelling by public transport and pedestrians.

Therefore, all these benefits are part of the challenges that the intelligent mobility in the Smart City idea has to face with regard to road transport [13].

In summary, observations of market reality indicate that few cities have a transport system that corresponds to the aforementioned solutions. The individual metropolises are implementing and using some elements of the smart transport system, but to a degree insufficient to recognize them as smart in this area. On the basis of the considerations presented above, the author puts forward the thesis that the city, in order to be perceived as "smart", should first of all try to use the services in the field of intelligent mobility that are currently available on the market.

3 Possible Technical and Technological Solutions in the Field of Intelligent Mobility in Road Transport

3.1 Autonomy of a Vehicle

The available literature provides a broad description of technological solutions supporting the so-called "vehicle intelligence", therefore the author in his further considerations will focus on the description of these services, which in the future are to be crucial for the implementation of the project of intelligent mobility. One of such innovative solutions is the "concept of an autonomous vehicle" [14]. The use of the GPS system and other sensors comes to limiting and eliminating the human factor, in order to improve the safety and efficiency of the means of transport, that is the car. All activities related to "steering" and other operations are performed without human interactions, thus extending the technical and economic characteristics and parameters of the vehicle [15]. Thanks to the use of navigation systems supported by traffic data, the optimal route for individual destinations is selected. From the information contained in the digital maps and the current position, the system calculates the most economical route to the destination point using, for example, a vehicle charging station. At the further stage of the development of the "autonomous vehicle" using additional thematic applications, it will be possible to model traffic in transport networks in urban areas in relation to both public, freight and individual transport [16].

Further increase of vehicle autonomy through intelligence, optimization and proper interaction with the environment will lead to increased road capacity and speed of travel. This will contribute to the growth of smart cities by optimizing transport. Additional benefits will be lower environmental pollution and energy consumption caused by increased traffic flow.

3.2 Intelligent Road

With the help of the latest generation of telematics solutions, the intelligent way not only informs about all kinds of dysfunctions, but also guides users to the goal. It is simply the way "alone" connects with telematic applications installed in the vehicle. On a current basis, the transport route measures traffic intensity, weather conditions, demand for information from the website. According to the observations of market reality, there is a tendency that more and more intersections in urban agglomerations gain traffic lights. In addition, pre-existing signaling is modernized in many places, as the city authorities strive to create an intelligent traffic control system. The aim of this system in the metropolitan transport system is to: improve the capacity of streets, (without expanding them), increase the level of security and improve public transport.

An important function of the system is to prioritize public transport priority, which is why the cycle of lights at many intersections depends on trams and buses. Information about vehicles going and waiting at the intersection, the computer controlling the duration of individual lights, is collected using appropriate detectors, induction coils and on-board computers installed in trams and buses.

The second parameter, which is worth mentioning are colloquially called "green waves". In large urban agglomerations, it is often the case that this is the case. the bedrooms are located outside the city center. The greatest traffic congestion occurs in the morning and afternoon hours. The solution to this problem is the intelligent way. The system analyzes the intensity of traffic depending on the time of day and at the right time creates the so-called green wave to unload the congestion problem as quickly as possible. Intelligent traffic lights are intended to improve street capacity. It is a fairly expensive way to improve communication in the city, it is more expensive than the development of cycling infrastructure, but much cheaper than street expansion or construction of multi-level intersections.

3.3 Positioning and Monitoring of the Vehicle

When discussing the subject, it is necessary to distinguish between vehicle positioning and monitoring. Positioning consists of reading the current position by the GPS receiver installed in the vehicle and passing this information to the destination point, e.g. to the transport base, the dispatcher or the insurance company. This process is not continuous, it can be it is random. While monitoring, in the simplest way, is based on cyclical reading of data coming from the means of transport about its position, archiving this data and transferring it to the interested parties using diagrams, maps, charts, etc.

Currently, vehicles travelling on the road are increasingly equipped with integrated systems, which include the GPS system positioner, transceiver system, memory and microprocessor control system. In addition, some of them allow you to connect the on-board computer keyboard to the printer, bar code readers, sensors and alarm systems.

The software adapter to the needs of a given vehicle is an integral part of this system. For example, another application will find its use in relation to public transport, i.e. urban buses and another one to freight transport. At this stage of the discussion, it should be mentioned that the positioning and monitoring service may include not only the means of transport, cargo, but even passengers.

In addition, positioning systems with the appropriate applications can also counteract any terrorist attacks aimed at taking over the vehicle (e.g. tractor with a semi-trailer), which due to their mobility and large loss of potential in relation to people and infrastructure becomes an easy target of the crime activity.

3.4 Control and Diagnostics of Technical Parameters of a Vehicle

In transport processes, diagnostics of means of transport is becoming more and more important in order to improve operational parameters and extend their "vitality".

This service allows to learn many important data and parameters directly during the operation process, including: fuel consumption, engine speed, oil pressure, speed, cargo space violations, cargo damage, alarm and fault analysis. Thanks to the use of additional sensors connected to the system, it is possible to request the authorization of the person trying to open individual cabin doors, or to monitor temperature, for example, in cold stores on a regular basis [13]. The service gives you the ability to control each vehicle subassembly after using the appropriate sensor. The monitoring systems can also work with anti-theft systems, which is associated with the increased security. The wide range of possibilities of diagnostic systems is currently a huge tool that can instantly transmit all data enabling full control both of the vehicle itself and its load, but also of the passengers.

3.5 Reconstruction of Dangerous Situations and Road Collisions

One of the priority demands regarding mobility within cities is to improve security. However, sometimes in the transport process, there is a situation when the vehicle or cargo is damaged, sometimes even the vehicle's passengers lose their lives. The problem is to determine the responsibility, causes and reconstruction of events. In these cases, experts in the field of road accident reconstruction will be able to reach for services in the field of smart mobility [16].

As the observations of market reality indicate, monitoring and delivered data sent in a cyclical manner are not sufficient to make a proper analysis and issue opinions in the analysed case. What can also have a significant impact on costs, especially when it comes to the cessation of the cause as a result of which damage of the property of a very high value occurred or the natural environment was polluted. By combining thematic applications of positioning systems and black boxes, which become standard in most motor vehicles, these situations can be reproduced with extraordinary accuracy. The process itself can be tracked repeatedly, not only in real time, but also after many months. It should be expected that this service will become a helpful "weapon" in this new and growing field of knowledge.

3.6 Road Rescue Services

The eCall system is a key example of the use of smart transport systems in relations to road transport. It is an automatic communication system in a crisis situation in the relationship: vehicle service, service – emergency services, vehicle – emergency services. The system architecture now includes the available satellite systems that increase the security levels in the event of a sudden call for support services. Thanks to the use of positioning systems, all data regarding the vehicle is transferred, i.e. the name, model, type, position, UTC time, type of danger and expected help. The main advantage of the system is the possibility of immediate notification of the danger and the launch of a quick rescue operation, even if such a notification is not made by the driver himself [16].

3.7 Intelligent Parking Systems

The exemplary intelligent parking system is based on a special application that is configured with the vehicle's communication system. This application is relatively easy to use and provides information about free places in parking lots and their exact locations. It can even lead the user to a given place. Importantly, the application recognizes the size of the car, preferred location, type of space, etc. The system notifies the user of the vehicle when their vehicle moves and notifies the service. In addition, it responds to bumps or emergency lights. For this purpose, several dozen or several hundred cameras are installed, whose image is analyzed by special algorithms.

3.8 Intelligent Toll Collection Systems

The system's operation consists in the camera (mounted on a pole or crane on the road) taking pictures of the vehicle's license plate as it passes through the detection point. Software using Optical Character Recognition (OCR) reads the license plate. The system then checks the recognized table in the database to identify the vehicle and determine the fee that the vehicle owner or driver has to pay. The advantage of the system is that there is no need to install any on-board device. The fee is charged to the account registered in the system. The toll collection systems can be based on technologies such as: intelligent video monitoring, license plate recognition or RFID (radio frequency identification), GPS or GSM system. Technically, these solutions can be adapted to charge a fee in advance or after using a given road. Additionally, the toll collection system can be integrated with databases.

4 Summary

As indicated by the analysis of available literature, urbanization is one of the most important social processes taking place on all continents, characteristics for our times. Urbanization defined as a process of concentration of people in specific points of geographical space, mainly in urban areas. It means a dynamic increase in the number of urban residents and their share in the population of a given area, and thus the increase in demand not only for transport services but also for the means of transport itself. Observations of market reality indicate that the current technical and technological progress and innovations within telematics applications create new opportunities to meet the transport needs of urban residents, which in turn may translate into improvement of their quality of life, identifying this fact with the development of the city.

The available literature emphasizes that in smart cities, the greatest developmental benefits can be achieved thanks to the integration of all transport branches affecting sustainable development [4, 6]. I am talking here, among others, about the effective management of urban transport, sustainable and ecological transport, elimination of congestion, or improvement of road users safety. Therefore, the main goal in the field of mobility is to create a sustainable, open and effective mobility system for both goods and people. According to many experts, the response to these needs may be the services offered by intelligent transport systems. Because the idea of intelligent mobility is based on strategic planning intentions governing public and private transport, changing the approach to traffic management and communication infrastructure of cities.

The authors agrees with the postulate promoted in the literature on the subject that in the coming years there will be a further dynamic development of new technologies in road transport, in particular in the field of information and telecommunications techniques [14–16]. This process will be of a global nature. The access to relevant information, and above all the speed of its processing and transmission will become one of the most effective mobility management tools, not only for the individual, but also for the collective one [17]. On this basis, the authors conclude that the use of intelligent transport systems in road transport creates new ways of managing transport activities, and thus leads to an increase in the importance of this mode of transport in relation to others. This means that vehicle manufacturers, meeting the expectations of customers, are already increasing the level and range of services available in the field of intelligent mobility. Therefore, emphasizing that a smart city cannot exist without intelligent mobility within road transport [18].

It is worth noting that the European Union also recognized the benefits associated with the implementation of intelligent transport solutions throughout min. in cities. In 2019, the European Parliament's Committee on Internal Market and Consumer Protection (IMCO) voted in favor of a major reform of vehicle safety standards. From 2021, each new vehicle will have systems already known from vehicles from the segment of luxury cars, for example: intelligent speed assistant, fatigue and driver distraction systems, lane maintenance assistant, tire pressure monitoring system, advanced emergency braking system, black box (event data recorder) that would help recreate the circumstances of the accident. It is also planned to equip all public transport vehicles with advanced systems for monitoring the driver's work and focus on the road as well as improved pedestrian and cyclist recognition systems.

However, this does not change the fact that as part of the development of the Smart city concept, there are several issues in this intelligent transport, for which, as of today, there is no answer in the literature on the subject:

- What will be the actual length of roads in cities where it will be possible to use intelligent mobility solutions, e.g. for autonomous vehicles?
- How will economic estimates change with the market's saturation rate with smart mobility solutions in cities?
- How will economic estimates change depending on, for example, the number of implemented intelligent transport solutions? Will the effects be visible before fully autonomous vehicles appear?
- Is the introduction of smart mobility solutions a higher priority than increasing the average speed of travel?
- How to ensure cost-effectiveness for intelligent transport solutions in the Smart City concept?
- The costs of implementing smart mobility solutions in given cities outweigh the possible benefits. In particular, where the infrastructure will require significant financial outlay and interest in, for example, autonomous vehicles will be small?

- Will the availability of intelligent transport systems, e.g. in 15 years, still be one of the postulates for the development of the Smart City idea?

Summing up the considerations undertaken by the authors regarding the accessibility to intelligent transport systems as the development of the smart city ideas, they do not fully exhaust the essence of the issue, but are only an attempt to signal the complexity of the presented problem. While the correctness of the proposed assumptions will certainly be verified by the market within a few years, which will allow further assessment of the impact of intelligent transport systems on the implementation of the Smart City concept.

References

1. Albino, V., Berardi, U., Dangelico, R.M.: Smart cities: definitions, dimensions, performance, and initiatives. J. Urban Technol. **22**(1), 3–21 (2015)
2. Correia, L.M.: Smart cities applications and requirements White Paper, Net! Works European Technology Platform, pp. 10 (2018)
3. Lombardi, P., Giordano, S., Farouh, H., Yousef, W.: Modelling the smart city performance, innovation. Eur. J. Soc. Sci. Res. **25**(2), 137–149 (2012)
4. Giffinger, R.: Smart Cities Ranking of European Medium-Sized Cites, Centre of Regional Science (SRF) Vienna University of Technology, pp. 45–46. Vienna Austria (2007)
5. Nam, T., Pardo, T.A.: Smart city as urban innovation: focusing on management, policy, and context. In: Proceedings of the 5th International Conference on Theory and Practice of Electronic Governance, September 26–28, pp. 56–57 Tallin (2011)
6. Woods, E.: Smart cities. Infrastructure, Information, and Communication Technologies for Energy, Transportation, Buildings, and Government: City and Supplier Profiles, Market Analysis, and Forecasts, Pike Research (2013)
7. Greenfield, A.: Against the Smart City New York: Do Projects (2013)
8. Colino, G, Rizzo, F.: A human smart cities. Rethinking the Interplay between Design and Planning, Springer International Publish, pp. 20–25. Switzerland (2016)
9. Zygiaris, S.: Smart city reference model: assisting planners to conceptualize the building of smart city innovation ecosystems. J. Knowl. Econ. **4**(2), 217–231 (2013)
10. World Economic and Social Survey 2013 - Suslainahle Derelopment Challenges. Annual report, 2013, The Department of Economic and Social Affairsofthe United Nations (DESA). https://sustainabledevelopmcnt.un.org/content/documents/2843WESS2013.pdf
11. World Urbanization Prospects: The 2011 Revision, 2012, United Nations Department of Economic and Social Affairs/Population Division, August. http://esa.un.org/unup/pdf/FINAL-FINAL_REPORT%20WUP2011_Annextables_01Aug2012_Final.pdf
12. Eskandarian, A.: Introduction to intelligent vehicles. In: Eskandarian A. (eds) Handbook of Intelligent Vehicles, pp. 1–13. Springer, London (2012)
13. Sun, Z., Bebis, G., Miller, R.: On-road vehicle detection: a review. IEEE Trans. Pattern Anal. Mach. Intell. **28**(5), 694–711 (2006)
14. Bisnath, S.B.: Precise Orbit Determination of the Champ Satellite with Standalone GPS 2012, pp. 34–35 Berlin (2010)
15. Autonomous Cars: Self Driving the New Auto Industry Paradigm, Morgan Stanley Blue Paper Reportm Morgan Stanley, pp. 15–16. New York City (2013)

16. Branco, P., Santos, R.: Intelligent car navigation systems – technology preview and future European directions. In: The European Navigation Conference GNSS 2004, pp. 18–19. Rotterdam (2004)
17. Zhu, F., Ukkusuri, S.: Modeling the proactive driving behavior of connected vehicles: a cell-based simulation approach. Comput.-Aided Civil Infrastruct. Eng. **33**(4), 262–281 (2018)
18. Fernandez-Anez, V.: Stakeholders approach to smart cities: a survey on smart city definitions. In: Alba, E., Chicano, F., Luque, G. (eds.) Smart Cities. Smart-CT 2016, Lecture Notes in Computer Science, 9704, pp. 157–167. Springer, Cham (2016)

How to Level the Playing Field for Ride-Hailing and Taxis

Kaan Yıldızgöz[1,2(✉)] and Hüseyin Murat Çelik[3]

[1] International Association of Public Transport-UITP, Brussels, Belgium
kaan.yildizgoz@uitp.org
[2] Institute of Science and Technology, Istanbul Sabahattin Zaim University, Istanbul, Turkey
[3] Istanbul Technical University, Istanbul, Turkey
celikhus@itu.edu.tr

Abstract. Introduction of ride-hailing services created many legal challenges in various countries because of their different features of operation and business model. Policy makers and regulators are pressured to define methods to handle these challenges. Today different countries has taken different approaches as policy and regulatory framework response to the entrance of ride-hailing companies but there is no common approach agreed. This research is meant to provide a system analysis and structured assessment regarding regulatory conflicts after introduction of ride-hailing services in different cities and also regulatory responses of these cities. This study presents an analysis and evaluation of developments to transportation policy and regulation since the proliferation of ride-hailing services with various examples and provide how to effectively approach "leveling the playing field" between traditional taxis and ride-hailing with set of principles to be considered while defining regulatory approach.

Keywords: Ride-hailing · Taxi · Regulation · Shared mobility · Governance · Technology

1 Introduction

One of the biggest changes in urban mobility during last 5 years is emergence of ride-hailing companies. They emerged during last 5 years and today completely shaping the urban mobility all over the world with many services in the spectrum of shared mobility modes. Ride-hailing companies are the platforms operated through mobile applications that match a customers demand for a ride with drivers. There are today different terminologies used for this type of services and platforms. Transport Network Companies, ride-selling and ride-sourcing are the other terminologies commonly used and today the best known example of such platforms is Uber, which launched its service in 2009. Since Uber's launch, several other companies have copied its business model like Lyft in USA, Ola Cabs in India, Didi in China, Careem in Middle East and Grab in South East Asia are some of examples [1].

Ride-hailing companies are facing different challenges in different countries. They were protested in many countries by traditional taxi industry and main discussions were about being against to fair competition rules, not paying taxes, driver qualifications and licenses, background check, passenger safety, insurance, non-compliance land passenger transport regulations. But they were appreciated by customers as experience and better service in comparison to traditional taxi operators. It is also important to consider impact of ride-hailing services to traffic congestion where there are different opinions claiming complementarity of ride-hailing services to mass transit or as opposite, Ride-hailing services impact for reducing mass transit ridership and increase of traffic congestion. Regulation of Ride-hailing services is very complex issue under the light of all points mentioned above. It is being discussed in some studies but not addressed in details until now. According to Geradini [2] taxis and ride-hailing should be regulated with separate framework as because the services proposed by taxi companies and online-enabled car transportation services are currently so far apart, it may be difficult to find a regulatory regime suiting them both. While incumbent taxi companies may wish to ensure that ride-hailing companies are forced to comply with the same regulatory requirements as applying to them, such an approach is a non-starter for ride-hailing as it would eviscerate their business model. McBride suggests in his thesis that the key aspects of the basic economic model, price and supply, should remain unregulated. As individuals familiarize themselves with the model, consumers will become more educated when making consumption choices. Therefore, it should not be necessary for legislative action to regulate ride-hailing pricing or supply as suggested by McBride [3]. Buckley [4] thinks the cities should follow the way of 'self-regulation' for ride-hailing services. Instead of applying taxi regulations to ride-hailing, a new form of regulations should be established for this market: self-regulation. As stated by Sundararajan self-regulation is not the same as deregulation or no regulation. Instead, public policy makers shift the onus from government and the public sector to the industry [5]. According to Merkert, since Uber simply assists in connecting drivers and riders, and does not own any vehicles that provide transportation; the company argues that it should not be governed by regulations concerning taxis. It is evident that imposing regulations on Uber is not only against the antitrust laws, but these regulations will hurt consumers the most [6]. According to Wagner [7] while the private sector is quickly adjusting to the proliferation of ride-hailing services, the public sector has been hesitant to adapt to the existence of Uber and its competitors. Compared to taxis, ride-hailing companies operate virtually unregulated, avoiding licensing costs, driver insurance requirements, standard employee training and background checks, state-controlled fares, and fleet size caps. Report of Pew Research Center asked the opinion of US citizens about regulation of ride-hailing services [8]. According to the report most of ride-hailing users are aware of the debate over how best to regulate these services – and these users feel strongly that ride-hailing services should not be required to follow the same rules and regulations as incumbent taxi operators.

As it can be predicted, this topic will remain as essential issue of debate in the future. Today different countries has taken different approaches as policy and regulatory framework response to the introduction of ride-hailing services but there is no common approach. This research is meant to provide a system analysis and structured assessment regarding regulatory conflicts after introduction of ride-hailing services in different cities and regulatory responses of these cities. This study presents an analysis and evaluation of developments to transportation policy and regulation since the proliferation of ride-hailing services with various examples and provide how to effectively approach "leveling the playing field" between traditional taxis and ride-hailing with set of principles to be considered while defining regulatory approach.

2 Methodology

This paper will be mainly based on survey for urban transport authorities, focus group meeting and literature review for collection of information from different sources including press releases and secondary data sources.

2.1 Literature Review - Case Study Collection

The literature study aimed to give an up-to-date and structured overview of the literature in the field of ride-hailing and urban mobility. A systematic literature search was performed in a structured way based on screening and selection of relevant studies from databases available to the author (SCOPUS, Science Direct, Google Scholar, UITP) and accompanying snowballing as well as focus groups lead to the majority of usable literature. In addition, collection of information from different sources including press releases and newspapers conducted as ride-hailing is new concept and there is limited academic work in this area.

2.2 Survey for Urban Transport Authorities

Set of questions developed related regulation and impact of ride-hailing companies to urban transport and the survey was conducted to different cities with different size, geographical background, different mix of mode share. Survey was conducted via e-mail for the urban transport authorities of that cities which can present the full overview and the data was collected from that cities based on their own research and studies (e.g. travel survey, urban mobility master plans, other studies). Collected data went undergo internal (compatibility with other data from the same city) and external (compatibility with data from other cities) checks and adjustments.

20 cities from different regions of the world (Europe, Asia, North America and Middle East) were included into survey and 5 questions were asked are Abu Dhabi, Ankara, Brussels, Budapest, Dubai, Dublin, Frankfurt, Helsinki, Hong Kong, Kuala Lumpur, Lagos, Lisbon, Milan, Montreal, Moscow, Oslo, Phoenix, Prague, Singapore and Tehran.

Question 1: Did you launch any specific regulations for Ride-hailing or modify the existing regulation?

5 of 20 cities launched specific regulation or modified the existing regulation after for ride-hailing services. These cities includes Dubai, Kuala Lumpur, Montreal, Phoenix and Singapore.

Question 2: Do you regulate the fare level of ride-hailing services?

Only in Montreal there is specific regulatory approach to fare-levels of ride-hailing companies. In other cities either they don't regulate fare level of ride-hailing services or they are subject to same regulatory approach of traditional taxis.

Question 3: Do you measure the impact of ride-hailing services on Traditional Taxi Market?

Only Kuala Lumpur and Montreal are the cities expressed that they are measuring the impact of ride-hailing services on traditional taxi market.

Question 4: Do you measure the impact of Ride-hailing services on Public Transport (Metro, buses etc.)?

None of the cities joined the survey expressed that they are measuring the impact of ride-hailing services on public transportation.

Question 5: Did you see any impact of Ride-hailing services on traffic congestion in your city?

Oslo and Milan expressed that they noticed increase of traffic congestion because of introduction of ride-hailing services in their cities. Other cities didn't measure any change.

2.3 Focus Group Meeting

Focus Group meeting was organized at the occasion of UITP Training Programme on Transport Network Companies in Brussels in 3–5 September 2018. There were 20 participants from different countries and from different type of organizations including transport authorities, taxi operators, ride-hailing companies and international associations. Participants were given set of questions related ride-hailing regulation and taxis. They were provided a time to discuss among each other and make proper research. Group presented outcome of their discussions related questions and their findings were used as one of inputs to the assessment done in this research.

- How urban transport authorities should define their policies related to Ride-hailing companies?
- Please evaluate and discuss 4 categories (Innovators/adapters/protectors/Regulators) given above and discuss on their pros/cons from perspectives of:
  – Sustainable urban mobility and traffic congestion
  – Customer satisfaction
  – Impact on Traditional Taxis
- Please propose policy & regulatory framework for the future of Ride-hailing companies & Traditional Taxi Sector
- Please also consider new mobility services like MaaS, Autonomous Mobility and suggest role for Ride-hailing and Traditional Taxis in the tomorrow's Mobility Ecosystem.

3 Regulatory Conflicts Related Ride-Hailing Services

Most controversial issues discussed today related to ride-hailing services are using private vehicles for operation and unprofessional drivers, surge charging, sharing the taxis with others and data privacy.

3.1 Using Private Vehicles as Taxi and Unprofessional Drivers

Ride-hailing companies are using the private vehicles and unprofessional drivers for the operation for most of their service types. UberPoP is the most well known in this area but this is the case also for many other ride-hailing companies as well. Of course using unprofessional driver and regular private vehicles for such services decrease the cost of service and brings planning efficiency but there are concerns from many aspects of safety, security and quality. In addition, the impact of such service from perspective of fair playing ground with the heavily regulated other traditional taxi operators is also crucial point to consider [9].

3.2 Selling Available Seats in Vehicles During the Same Trip

Sharing the taxi trips with other passengers are not legal in many countries. There are different reasons behind the decisions taken by authorities for that position but today with applications like UberPool, Olashare this became commonly practiced in cities.

3.3 Surge Pricing

Of course surge pricing was at everyone's lips when it was first introduced by ride-hailing companies. Surge Pricing is a tool to match demand and supply for taxi rides when there is high demand. Technology allows changes in fare in dynamic environment based on complex algorithm. It is like pricing model of airline companies during peak seasons and off peak seasons by applying this to daily taxi trips was not usual initially. Surge pricing increases the fare by a given multiple. Lyft's peak time pricing algorithm will increase fares from 25 to 200% according to changes in supply and demand. Similarly, Uber's surge pricing can increase fares anywhere from a 1.2 times multiple to 7 times. Both firms' pricing algorithms automatically go into effect when potential consumers outpace potential drivers. Off-duty drivers receive real time updates regarding surge pricing as means to incentivize them onto the road or into certain neighborhoods [3]. According to Matthew Daus one result of Ride-hailing "surge pricing" is that communities with limited or no ride-hailing access, such as low-income and minority communities, may be "redlined" since drivers may choose not to operate in those areas [10].

3.4 Data Privacy

Tracking the users after the end of the journey by ride-hailing companies is also an important discussion mainly from the privacy perspective. Tracking generally happens from the moment customer requests a trip and after five minutes the journey had ended. Claimed that the change would improve the app by allowing for more reliable pick-ups, improving customer service, and enhancing safety. Apps are also tracking commuters social behaviors and to understand the purpose of journeys.

3.5 Taxation

Another area of discussion related ride-hailing companies is the place and the amount of tax paid by them. For example, It is claimed that Uber paid only 411.000 Sterlin as tax in UK in 2016 although the total turnover in this country was 23.3 million Sterlin and net profit was 1.3 million Sterlin. It is claimed that Uber established companies in countries where there is no or very less tax ratios applied and most of the revenues of the company is recorded in that countries [11]. According to Kuneshegaran Uber transfers the commission income from different countries to Uber BV which is established in Netherlands which is linked to Uber International BV established in Bermuda. This helps Uber to keep the amount of tax paid low in different countries [12]. Such tax optimization approaches are being used by many multinational companies as well.

3.6 Contractual Relationships with Drivers

Ride-hailing companies establish subcontractor relationship with their drivers via service contract signed between parties instead employing them as staff in their companies. This created discussion in many countries. In UK Labor Court decided in 26 October 2016 that Uber drivers have rights of minimum salary and annual paid leave [13]. In March 2018, It was decided by Swiss State Economic Affairs Secretariat that Uber drivers should be treated as employees [14].

4 Examples of Regulatory Practices

Practical approach in different countries for regulation of ride-hailing services also varies and they were analyzed for 25 countries below. Countries were selected to have balance between them in terms of different type of regulatory framework, geography and income level.

4.1 USA, Canada and United Kingdom

USA is the birth place of Ride-hailing disruption and it is of course where the most discussion happened until now. Regulatory response was different in different states of USA and also it has changed during last years as some states changed their policy and regulatory position on this issue. As of January 2017, 34 states and more than 69 cities have passed legislation governing ride-hailing companies (transportation network companies). Another six states have enacted legislation mandating minimum insurance requirements [10]. In New York, Taxi and Limousine Commission created Licensing & Rules for Providers of E-Hail Applications to regulate the service. In San Francisco, the authorities passed a bill regulating insurance coverage [15]. In Texas, the authority requires state-controlled fingerprint checks of drivers - performed by a vendor. Ride-hailing companies do not agree with the condition [16]. In Las Vegas, the city council requires each ride-hailing company to pay a flat annual US$17,500 license fee. Earlier, the council was charging per-driver fee [17]. Four cities In Missouri

(Springfield, St. Louis, Kansas City and Columbia) has legalized Ride-hailing services. Springfield City Council cleared the bill in November 2016, which requires Ride-hailing companies to do background checks. In Maryland, City Public Service Commission announced new guidelines for Ride-hailing services, which require fingerprint background checks of drivers [18]. As of mid-2016, Ride-hailing services were available in 14 cities in Canada – Calgary, Edmonton, Toronto/GTA, Ottawa, Montreal, Quebec City, London, Guelph, Waterloo, Kitchener, Niagara, Windsor, Hamilton, Kingston, covering 50% population of the country [19]. Edmonton became the first city to legalize Private Transport Providers and has passed bylaw. From March 1, 2016 ride-hailing companies must have provincially approved insurance, annual vehicle inspection, criminal record check and charge a minimum of $3.25 per ride, but there is no price cap. Only taxis are permitted to pick up street hails or use taxi stands [20]. Hamilton passed a bylaws in January 2017 which requires ride-hailing companies, with more than 100 vehicles, to pay $50,000 annual fee plus six cents per trip. Drivers are not allowed to pick up roadside clients [21]. London is the capital and biggest city of UK and also important benchmark always for taxi services. Transport for London (TfL) released new regulation for private-hire drives in June 2016. The key condition includes - English language requirement for drivers, accurate fare estimates, Panic response, Driver & vehicle details before the journey, and 'Hire and reward' insurance requirements. London City Council passed the proposal which will require drivers to install cameras in ride-sharing vehicles. Further, it has proposed to charge 26 cent per ride fee to drivers. An employment tribunal in London ruled its licensed drivers should be classed as workers with access to the minimum wage, sick pay and paid holidays. TfL also didn't renew the license of Uber with the decision end of 2017 [22]. United Kingdom was grouped together with USA and Canada but not together with other EU Countries with considering Brexit process.

4.2 Brazil and Mexico

In Brazil, The federal government is working on national guidelines for Transport Network Companies but there are different initiatives of cities already taken in this regard. In Sao Paulo, the mayor passed the regulation to allow the use of Transport Network Companies. As per the regulation, the city shall charge an average fee of 10 centavos ($0.03) a kilometre for drivers working with TNCs. The fund are collected into a municipal fund. Mayor of Rio de Janeiro signed a law forbidding private cars to transport passengers in November 2016. Only registered taxis are allowed in the city. However, this not impact Uber operation as federal guidelines are awaited [17]. In Mexico city, the city council issue regulation for taxi companies, which includes a 1.5% ride levy on the cost of each trips, an annual permit fee and the establishment of a minimum value for each vehicle. The levy is deposited into a new fund for Taxis, Transportation and Pedestrians and will be used for investments in better mobility options for the city. The city council of Tijuana declared legal in July 2016. The operators will require to pay 1.5% of gross annual turnover. Further, vehicles must not be longer than six years old and a minimum value of US$8,000, having insurance for US$ 150,000 to cover damage to third parties and passengers [23].

4.3 United Arab Emirates and South Africa

Dubai become the first city to regulate ride-hailing services by entering into agreement with Careem and Uber. Ride-hailing apps will offer luxurious transport services (limousines) via online and smart Apps channels. All cars used by them must belong to accredited companies. The laws requires that ride-hailing apps must charge at least 30% higher fares than taxi fares. Both Careem and Uber stopped service in Abu Dhabi in August 2016 owing to regulatory issues. However, Careem has resumed service in February 2017 under the regulations for limo services [1]. South Africa's National Land Transport Act (NLTA) only recognizes six categories of private transport services: buses, minibus taxis, metered-taxis, chartered vehicles, lift clubs and tuk-tuks (motorized rickshaws). South Africa has no regulations governing e-hailing services. The government amended NLTA and included a sub-category to accommodate e-hailing services. As per the amendment, e-hailing service operators are classified as metered taxi operators. The drivers require to have metered licences as part of the law. The legislation allows the use of a smartphone in lieu of a taximeter, and requires the operators to estimate distance and fare in advance, as well as, provides driver details [24].

4.4 Australia and New Zealand

Australian Capital Territory (ACT) legalized ride-hailing services in October 2015 and issued new regulation, which include background checks, vehicle inspections and insurance requirements for drivers. New South Wales legalized ride-hailing services in December 2015. The law requires criminal and car-safety checks. The authority established A$250 million "industry adjustment package" to compensate taxi drivers [24]. South Australia legalized Uber like services in July 2016. Ride-selling apps requires to follow safety standards prescribed in Passenger Transport Act. Taxi services were offered compensation including A$30,000 per licence and all metro taxi trips will have a A$1 levy to fund this assistance [25]. Queensland legalized ride-booking services in September 2016. Initially, there will be no license requirement for the operator. However, the authority will introduce new license system in 2017. As part of the regulatory changes, there will be a A$100 million assistance package for the taxi industry, which includes A$20,000 one-off payments for taxi licence holders and A$10,000 per limousine licence. Western Australia is looking to overhaul and deregulate taxi industry. The plan includes a A$27.5 million "transition assistance package", including compensation payments of A$20,000 for taxi plate owners [26]. New Zealand The Ministry of Transport issued guidelines for Small Passenger Services (SPS) Regulations. The revised bill proposed to bring taxis, private hire services such as limousines, shuttles, ridesharing, and dial-a-driver services into a single category. Taxis will continue to be defined as a SPS and will operate in much the same way as they were doing earlier. A technology or app based operator is also defined as a SPS, and therefore will be required to become an approved transport operator (ATO). The bill proposes many concession for ridesharing companies including allowing drivers to get the required background and compliance checks easily, as well as, not required to pass area knowledge and English language tests. Further, it scrapped safety requirements such as in-taxi cameras for ridesharing services. The bill requires ridesharing companies to present vehicles for inspection "at a moment's notice", and to keep fuel receipts and collect logbooks from its drivers [27].

4.5 Philippines, Singapore and Malaysia

Philippines was the first country in South-east Asia to legalize TNCs in May 2015. Department of Transportation and Communications (DOTC) created new categories of Transportation Network Vehicle Service (TNVS) to allow app-based services offered by TNCs to exist within our regulatory framework. Under the new classification, a TNC is defined as an organization that provides pre-arranged transportation services for compensation using an internet-based technology application or a digital platform technology to connect passengers with drivers using their personal vehicles. TNCs will provide the public with online-enabled transportation services known as a TNVS, which will connect drivers with ride-seekers through an app [28]. Singapore is one of the most important cities when we are talking about taxi services because of its unique model of taxi regulation and efficient taxi operations. Land Transport Authority of Singapore has introduced "Third Party Taxi Booking Service Providers Act" with effect from 1 September 2015. All third party taxi service providers with more than 20 participating taxis are required to be registered with LTA in order to operate the booking service [29]. SPAD, Nationwide Land Transport Authority of Malaysia, has issued Taxi Industry Transformation Programme (TITP) to reform taxi industry, including legalizing e-hailing operators which are mainly Uber and Grab in the country [30].

4.6 Japan, South Kore and Taiwan

There is no separate regulation for ride-hailing companies in Japan. The operators need to follow the same rules and regulations, applicable for conventional taxis [31]. In Seoul, the government banned Uber and other App based companies. Seoul Metropolitan Government launched its own taxi app for registered cabs. It has currently allow KakaoTaxi to provide taxi booking service [32]. As per Taiwanese Laws, taxi companies must be domestically owned and operated. The government has refused to propose a separate act for technology firms. Uber is facing lots of problem in Taiwan since the commencement of operation in 2013. Uber is registered in Taiwan as an information management company as it allows an foreign company to operate in Taiwan, but the government sees it as a transportation services company. The company stopped the operation in February 2017 [33].

4.7 Russia, China, India and Vietnam

There is a federal law in Russia which is very liberal and permissive for taxis. Actually, any person or company can become a taxi driver or operator. There is no restriction at all except the age of the driver (min 21 years old), no test, no medical control, no criminal record, no license fee, no restriction on the type or age of the car, no obligation to have a taxi meter, no regulation of fares, etc. Anyone can make a request online and will get the license. There is no ceiling on the numbers of taxis in the city. The only condition is to paint the car in yellow and put a taxi sign. Currently, 72% of taxis are yellow colors. The drivers have to pay a tax equal to 6% of their theoretical revenue to the department of transport. Taxi-hailing apps can only work with license drivers and require sharing data with the authorities. In July 2016, the government has imposed 8% value-added tax on electronic goods and services provided by global internet giants including Uber starting in 2017. Foreign companies have no obligation to create a legal entity in Russia and can pay the tax via local partners. Uber has directed its Russian drivers be registered as legal entities or individual entrepreneurs, and that the responsibility for the settlement of the tax be specified in their contract with Uber N.V., which is registered in the Netherlands [34]. The Chinese government released the Interim Measures for the Administration of Online Car-hailing Operations and Services in July 2016. The policy was released by the Ministry of Transport and six other ministries jointly. This is the first national measures that legitimize online car-hailing service. The new regulations became effective on 1 November 2016. As per the guidelines, transport authorities under the State Council is responsible for guiding the national network about car management [17]. Indian Ministry of Transport issued the new policy for taxis on 15 December 2016 and circulated the same to all cities/states [17]. In April 2014, Vietnam government released the guidelines for ride-selling application. As per the guidelines, technology companies required to sign contracts exclusively with local commercial transport companies that comply with regulations such as identifying cars with official registrations and logos, and equipping them with tracking devices. However, the implementation of the rules are still pending. The government is considering to revamp the policy [35].

4.8 EU Countries (Germany, France, Spain, Belgium, Denmark, Italy)

In Germany, local laws require taxi drivers to hold commercial licenses in order to pick-up passengers and adhere to a set fare structure. There is no separate regulation, so on-demand transport services need to comply with existing taxi laws [36]. French authorities earlier imposed the rule forcing Car services to wait for 15 min between reservation and pick-up. The government has merged "Collective Transport Permit" with "Chauffeurs License" to make it difficult to obtain license [37]. In Spain, On-demand transport services companies can only work with drivers who carry a valid professional VTC license, as required by all professional drivers [38]. In Belgium, On-demand transport service is banned in the country for using private cars. Only license taxis service is allowed, for example Uber drivers may be fined 10000 Euro for any pick-up [39]. Denmark has introduced new taxi laws in February 2017 that includes requirements such as mandatory fare meters, video surveillance and seat occupancy detectors to activate the airbags [40]. The Italian government has deferred the introduction of norms to control car-hire and car-share services till the end of 2017. The ride-hailing companies buy licenses in smaller towns where it cost less and use them to work in cities. A taxi license in Rome is worth EUR 150,000 but the NCC (cars rented with a driver) license just one-tenth [41]. For European countries, it is also important to mentioned European Court of Justice took a decision at the end of 2017 and ruled that Uber is a transport services company, requiring it to accept stricter regulation and licensing within the EU as a taxi operator [42].

5 Categorization of Regulatory Approaches

When we analyze the different legal and policy responses we can categorize them into 4 as 'Innovators' who are creating new regulation to integrate TNCs to urban mobility offer, 'Adapters' mainly following other countries and adapting themselves to new environment, 'Protectors' who are protecting existing legal approach and 'Regulators' that allow ride-hailing companies to work but with very stringent conditions. Comparison could be better seen in Regulatory Matrix provided in Fig. 1. Of course there could be different ways and parameters of categorization but is a simplified approach. Also it is not easy to put all into a single structure as of very different situation among countries. Each country should be analyzed carefully.

Fig. 1. Comparision and categorisation of regulatory approaches.

6 Recommendations

Introduction of ride-hailing services created many legal challenges because of their different features of operation and business model. Today different countries has taken different approaches as policy and regulatory framework response to the entrance of ride-hailing companies but there is no common approach. With considering current practices in various countries, impact of ride-hailing to urban mobility and under the light of future considerations below could be provided as set of key recommendations for regulators to consider when they are defining their policies and regulatory framework related to ride-hailing services:

- Authorities should not priotise ride-hailing or taxi services over public transport as high quality public transport is the only alternative able to fulfil the lion's share of trips by using a minimum of space.
- It is important to promote integration of ride-hailing and taxi services to public transport. Without public transport, other sustainable & innovative mobility services cannot offer an affordable alternative to car ownership
- Regulatory Framework should not have the aim of protecting any certain group but should be designed as a tool to implement public policies,

- Regulatory Framework for Taxis and Ride-hailing services should be developed together to make sure fair conditions,
- Advanced technology can be used to develop advanced regulatory and enforcement schemes,
- Integrate policies and regulation related ride-hailing into larger efforts to implement wider usage of new technologies and innovative applications. Cities can use technology revolution and ride-hailing companies for better implementation of demand management practices with tools like UBER POOL or OLA SHARE
- Regulatory Framework should less concentrate on Price and Quantity Regulation but concentrate more on Quality Regulation in terms of road usage, environmental issues, safety and availability.
- Regulatory Scheme based on Road Charging should be considered to avoid extra vehicle miles and congestion of ride-hailing services like the model Brazil. Road Charging could be applied as an incentive, for instance, there could be different pricing in terms of passenger occupancy in ride-hailing cars. Incentivized pricing could be applied to promote shared vehicles, if rides are shared no pricing implemented
- Authorities should promote partnerships between ride-hailing and traditional taxi actors. This option could bring benefit to both parties and they can act complementary other than competing each other and losing sources for this competition for the midterm.
- Cities mainly in emerging countries should consider using ride-hailing companies to help regarding rationalization and formalization of individually owned/operated taxis
- Regulation should not only concentrate competition between ride-hailing and taxis/public transport but also potential competition may take place soon in the future between Ride-hailing companies

7 Conclusion

Implementation of above mentioned recommendations will require more detailed guidance and understanding of local conditions and capacities, which is outside the scope of this study. Next steps could include developing this guidance, particularly for more comprehensive analyses and detailed guidelines. While not covered in the scope of this study, deeper analysis and understanding is needed of the impacts of ride-hailing services on employment market, especially in developing countries where driving for ride-hailing has made it easier for people to access the job market. It is also quite essential that evaluation of the results and impact of different policy actions and regulatory responses is needed to better understand which tools are effective and which are not. Finally, it is important to work related potential role of ride-hailing in the future of urban mobility with considering latest developments especially related autonomous vehicles and other shared mobility and technology innovations like MaaS-Mobility as a Service. This study didn't suggest an ecosystem and organizational structure related tomorrow's Sustainable Mobility Ecosystem with considering emergence of Transport Network Companies, Autonomous Vehicles and Mobility as a Service.

References

1. Yıldızgöz, K.: Dijitalleşme Çağında Taksiler, Marmara Kultur Yayınları. Istanbul, Turkey (2018). ISBN 9786056807169
2. Geradini, D.: Should Uber be Allowed to Compete in Europe? Legal Studies Research Paper Series, Gearge Mason University, USA (2015)
3. McBride, S.: Ridesourcing and the taxi marketplace. Electronic thesis or dissertation, Boston College USA (2015)
4. Buckley, C.: An Examination of Taxi Apps and Public Policy Regulation. http://clarebuckley.ca/pdf/Clare%20Buckley%20-%20public%20policy%20regulation.pdf
5. Sundararajan, A.: Peer-to-Peer Businesses and the Sharing (Collaborative) Economy: Overview, Economic Effects and Regulatory Issues. http://smallbusiness.house.gov/uploadedfiles/1-15-2014_revised_sundararajan_testimony.pdf
6. Merkert, E.: Antitrust vs. Monopoly: an Uber Disruption. FAU Undergraduate Law Journal. http://journals.fcla.edu/FAU_UndergraduateLawJournal/article/view/84609
7. Wagner, D.: Sustaining Uber: opportunities for electric vehicle integration. Pomona Senior Theses, University of Claremont, USA (2017)
8. Pew Research Center: Shared, Collaborative and On Demand: New Digital Economy. http://www.pewinternet.org/2016/05/19/the-new-digital-economy/
9. Yıldızgöz, K., Çelik, H.M.: Critical moment for taxi sector: what should be done by traditional taxi sector after the TNC disruption? In: Nathanail, E., Karakikes, I. (eds.) Data Analytics: Paving the Way to Sustainable Urban Mobility: CSUM 2018. Advances in Intelligent Systems and Computing, vol. 879. Springer, Cham (2019)
10. Daus, M.: The Expanding Transportation Network Company "Equity Gap". University Transportation Research Center, New York, USA (2016)
11. Bowers, S.: Uber's main UK business paid only £411,000 in tax last year. https://www.theguardian.com/business/2016/oct/10/ubers-main-uk-business-paid-only-411000-in-tax-last-year
12. Kunashegaran, S.: How Uber, Google, Facebook and Other Tech Giants Avoid Paying Billions in Tax? https://medium.com/dconstrct/how-uber-google-facebook-and-other-tech-giants-avoid-paying-billions-in-tax-365b7c8b7dbc
13. Griswold, A.: A British court rules Uber drivers have workers' rights in the "employment case of the decade". https://qz.com/822104/the-london-employment-tribunal-rules-uber-drivers-have-workers-rights-in-Europes-employment-case-of-the-decade/
14. Swissinfo,: Swiss authorities say Uber drivers should be treated as 'employees'. https://www.swissinfo.ch/eng/wage-dumping-_swiss-authorities-say-uber-drivers-should-be-treated-as–employees-/43984356
15. Public Utilities Comission, State of California: Basic Information for Passenger Carriers and Applicants. http://www.cpuc.ca.gov/uploadedFiles/CPUC_Public_Website/Content/Utilities_and_Industries/Passenger_Carriers_and_Movers/BasicInformationforPassengerCarriersandApplicants_Nov2014_11172014lct.pdf
16. Hawkins, A.: The five issues holding Uber and Lyft back in big states. https://www.theverge.com/2016/9/21/12987566/uber-lyft-five-issues-big-state-legislation-failure-schaller
17. Singh, J.: Combined Mobility and Public Transport, UITP Training Programme, Vienna, Austria (2017
18. Maryland Public Service Comission: Maryland PSC approves alternative backround checks for Uber and Lyft Drivers. http://www.psc.state.md.us/wp-content/uploads/MD-PSC-Decision-On-Uber-Lyft-Fingerprint-Waiver-Petitions_12222016.pdf

19. Licorish, D.: Ride-sharing in Canada: It's Complicated. http://www.lowestrates.ca/blog/ride-sharing-canada-its-complicated
20. Stolte, E.: Edmonton becomes first city in Canada to pass Uber friendly by-law. http://edmontonjournal.com/news/local-news/police-lift-lock-down-on-edmonton-council-chambers-for-uber-debate
21. Craggs, S.: Uber will be safer and more easily identified under new law: city. http://www.cbc.ca/news/canada/hamilton/licensing-uber-1.3942039
22. Chapman, H.: London taxi and PHVs. In: 4th International UITP Taxi Seminar, London, UK (2017)
23. Haldevang, M., Mexico to regulate Uber with licence fees, ride levy. http://www.reuters.com/article/us-mexico-uber-idUSKCN0PI17420150708
24. The Guardian: Canberra takes the lead in regulating Uber as NRMA urges others to follow suit. https://www.theguardian.com/technology/2015/sep/30/nrma-urges-governments-to-regulate-uber-and-other-ride-sharing-services
25. Fare: Official Journal of the Taxi Council South Australia. http://www.aitaxis.com.au/media/W1siZiIsIjIwMTYvMDYvMjgvNHRoZGZhbGQyX2ZhcmVfbWFnX2p1bmVfMjAxNl94Mi5wZGYiXV0/fare%20mag%20june%202016%20x2.pdf
26. ABC News, These are the states and territories where Uber is (and isn't) legal. http://www.abc.net.au/news/2016-08-11/where-is-uber-legal-in-australia/7719822
27. New Zealand Ministry of Transport. https://nzta.govt.nz/assets/Commercial-Driving/Small-passenger-services-transitional-guide.pdf
28. Land Transport Franhising and Regulatory Board: Rules and Regulations to Govern Transport Network Companies. http://ltfrb.gov.ph/media/downloadable/MC_NO._2015-015_.pdf
29. Teo, J.: Regulation of third party booking apps in Singapore. In: IATR Annual Congress, San Francisco, USA (2016)
30. Hassan, A.: Malaysia taxi transformation plan. In: UITP National Public Transport Conference, Kayseri, Turkey (2017)
31. Hanada, R., Urasaki, K.: Rules leave Uber with hard road in Japan. http://asia.nikkei.com/Business/Companies/Rules-leave-Uber-with-hard-road-in-Japan
32. Strange S.: Seoul bans Uber and plans to launch a competing app. http://mashable.com/2014/07/21/seoul-korea-bans-uber/#A6xwo.uj7PqQ
33. Mullen, J., Yang, Y.: Uber suspends its service in Taiwan as fines mount. http://money.cnn.com/2017/02/02/technology/uber-taiwan-suspending-service-fines/
34. Khrennikov, I.: Uber Russian Drivers Quit After Putin's Tax on U.S. Tech Giants. https://www.bloomberg.com/news/articles/2017-01-13/uber-russian-drivers-quit-after-putin-s-tax-on-u-s-tech-giants
35. Nikkei: Vietnam lays down law on Uber. http://asia.nikkei.com/Politics-Economy/Economy/Vietnam-lays-down-law-on-Uber
36. Davies, R.: Uber suffers leagl seatbacks in France and Germany. https://www.theguardian.com/technology/2016/jun/09/uber-suffers-legal-setbacks-in-france-and-germany
37. Dillet, R.: Uber, LeCab and others now have to wait 15 minutes before picking up you. https://techcrunch.com/2013/12/28/uber-lecab-and-others-now-have-to-wait-15-minutes-before-picking-you-up-in-france/
38. Che, C.: 9 countries that aren't giving an Uber inch. http://www.huffingtonpost.com/entry/uber-countries-governments-taxi-drivers_us_55bfa3a9e4b0d4f33a037a4b
39. Khan, J.: 10.000 Euro fines threat to Uber Taxis in Brussels. https://www.ft.com/content/b23e9ee4-c4b9-11e3-9aeb-00144feabdc0

40. Henley, J.: Uber shut down Denmark operation over new taxi laws. htttps://www.theguardian.com/technology/2017/mar/28/uber-to-shut-down-denmark-operation-over-new-taxi-laws
41. Heisler, Y.: Uber was just banned from operating in Italy. http://bgr.com/2017/04/08/uber-cars-banned-italy/
42. European Court of Justice: Asociacion Profesional Elite Taxi v Uber Systems Spain SL. https://curia.europa.eu/jcms/upload/docs/application/pdf/2017-12/cp170136en.pdf

Assessment of BEVs Use in Urban Environment

Ewelina Sendek-Matysiak^(✉)

Department of Mechatronics and Machine Construction,
Kielce University of Technology, Kielce, Poland
`esendek@tu.kielce.pl`

Abstract. Transport accounts for almost a quarter of Europe's greenhouse gas emissions and is the main contributor to air pollution in cities. Moreover, it is one of the most important noise sources shaping the acoustic climate in urban areas. For this reason, there have been actions implemented in the European Union for years now aimed at improving air quality, with a view to meeting by 2030 standards at the levels set by the WHO. One of them, which is to significantly reduce the environmental burden generated by road transport, is the replacement of conventionally powered cars with pure electric BEVs (Battery Electric Vehicle). This paper, basing on comparison of cars with different types of propulsion (gasoline, diesel, electric and plug-in hybrid engines), evaluates which of them is currently the most beneficial to use in the city in environmental, technical and economic terms. The analysis was carried out for two situations:

(1) total emissions from the production of fuel, manufacture of the vehicle and its use,
(2) vehicle emissions during its use
using one of the multi-criteria optimization methods – the point method.

Keywords: Electromobility · Noise · Point method · BEV

1 Introduction

On 16 May 2018, the Population Division of the United Nations Department of Economic and Social Affairs published a report entitled "Revision of World Urbanization Prospects" in which it stated that currently 55% of the world's population lives in urban areas, and this percentage will increase to 68% by 2050 (Fig. 1).

Forecasts indicate that urbanization, resulting from a progressive change of place of residence from rural to urban areas combined with a general increase in the world population, may result in an increase in the number of people living in urban areas by another 2.5 billion people by 2050 compared to 2018 (Fig. 2) [1]. It is estimated that in the European Union this number will increase by that time by 40,000 (Fig. 3), as a result of which the share of urban population in the total population will amount to about 80% (Fig. 1).

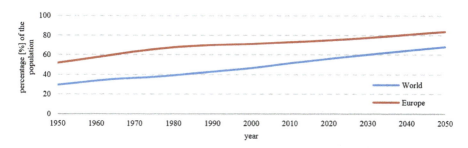

Fig. 1. Share of the population living in urbanized areas (own elaboration based on [1])

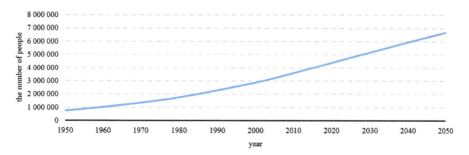

Fig. 2. Forecast of the number of people living in cities in the total world's population (own elaboration based on [1])

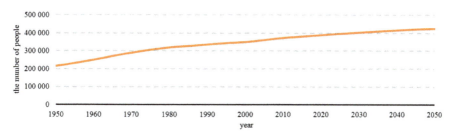

Fig. 3. Forecast of the urban population living in cities in the total population of the European Union (own elaboration based on [1])

A systematic increase in the level of urbanization has both positive and negative effects.

The concentration of consumers, workers and economic operators in one place or in the same area, combined with the formal and informal institutions that make an agglomeration 'dense' and coherent, fosters, among others, the realization of needs and aspirations related to the organization of social life, politics, the performance of economic tasks, the collection and dissemination of cultural material, to name but a few. As a consequence, cities, which are currently inhabited by more than 70% of the population of the European Union, contribute to the production of more than 80% of the Community's GDP [2].

However, the increasing degree of urbanization, apart from the positive effects, generates many negative phenomena, which are related to, among other things, transport activities based to a large extent on the use of passenger cars.

2 Effects of Vehicles Operation with Internal Combustion Engines in the Urban Environment

An extremely non-green and low-efficient internal combustion engine emits significant amounts of harmful combustion products into the atmosphere. These mainly include particulates (PM10 and PM2.5), nitrogen, sulfur, carbon and heavy metals, such as cadmium, lead and mercury oxides as well as unburned aromatic hydrocarbons.

The share of road transport by vehicles equipped with internal combustion engines in total air pollution and in cities is shown in Fig. 4.

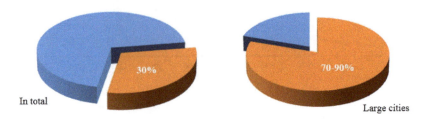

Fig. 4. The share of road transport in air pollution [3]

Products of fuel combustion in motor vehicles affect the natural environment, particularly on a local and regional scale, being the reason for the occurrence of such phenomena as acid rain or photochemical smog and shaping the microclimate around roads.

Harmful substances emitted to the atmosphere (e.g. suspended particulates, aromatic hydrocarbons, including benzo(a)pyrene, sulfur and nitrogen oxides) have serious health effects, including asthma, allergy, COPD (chronic obstructive pulmonary disease), fertility problems, preterm births, lower birth weight of children, lower IQ in subsequent generations, stroke, heart attack, ischaemic heart disease and even diabetes or Alzheimer's disease. Most of these conditions lead to premature deaths.

The WHO considers air pollution to be the greatest environmental threat to health in Europe [4, 5]. In the EU, air pollution causes on average more than 1,000 premature deaths every day, which is more than ten times the number of people killed in road accidents [6]. In addition, a further 6.5 million Europeans are diagnosed with serious respiratory and cardiovascular diseases (strokes, asthma, bronchitis). In Poland, 47.3 thousand people a year die prematurely because of air pollution [7]. In 2013 the EU Commission estimated the total external health-related costs of air pollution amount to between €330 and €940 billion per year [8].

According to the European Environment Agency, urban population is at the highest risk. In 2015, around a quarter of Europeans living in urban areas were exposed to levels of air pollution exceeding some EU air quality standards, and 96% of them were even exposed to levels of air pollution considered harmful by the WHO [9]. Air pollution generally affects urban dwellers to a greater extent than rural dwellers, as urban population density means that air pollution is released on a larger scale (e.g. from road transport) and dispersion is more difficult in cities than in rural areas.

In addition, transport is largely responsible for the deterioration of the 'acoustic climate' in urban areas.

According to the EU report [10], approximately 40% of the European population is exposed to traffic noise levels exceeding 55 db, while 20–30% to levels above 65 db during the day and 55 db at night.

Figures 5 and 6 show the approximate number of people exposed to harmful noise.

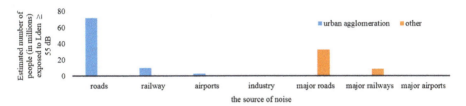

Fig. 5. Number of people exposed to day-time noise (Lden) $\geq$ 55 dB in EU-28 in 2017 [11]

Fig. 6. Number of people exposed to night-time noise (Lnight) $\geq$ 50 dB in EU-28 in 2017 [11]

At present, at least 10,000 Europeans die prematurely as a result of excessive noise and the costs resulting from the impact of road noise on public health are estimated at €40 billion per year [12].

In 1980, Directive 80/779/EC established limit values for SO2 in the EU for the very first time. In the following years, further directives on subsequent air pollutants and their limit values were adopted, which has consequently contributed to reducing air pollutant emissions in recent decades. Between 1990 and 2015, EU SOX emissions decreased by 89% and NOX emissions by 56%. Since 2000, PM2.5 emissions have been reduced by 26% [13]. However, the WHO and the EEA indicate that this reduction in total air pollutant emissions does not automatically translate into a similar reduction in air pollutant concentrations. EU legislation concerning pollution sources does not focus on reducing emissions where people suffer most from air pollution or where concentrations of pollutants are highest. For example, if car engines produce less pollution due to stricter emission standards, pollution may continue to increase if the level of car use increases. Therefore, specific action is needed in populated areas in order to reduce air pollutant concentrations, as human exposure, in particular to PM and NO2, remains high.

3 The Use of BEV Electric Cars in the City

An instrument for counteracting these negative phenomena is the intensively implemented strategy for the use of alternative vehicles in the European Union transport sector, including those purely battery-powered – BEVs. The scale of changes that are to take place in this field in the coming years is best illustrated by one of the assumptions of the White Paper entitled "Roadmap to a Single European Transport Area – Towards a competitive and resource efficient transport system" - halve the number of conventional cars in urban transport by 2030 and eliminate them from cities by 2050. Electric cars are those powered by electric energy stored in batteries. The drive system of a classic electric car consists of an electric power source, an electric machine (electric drive motor) and mechanical transmissions that transmit power to the driving wheels. The source of electric energy is an electric battery, while the proper operation of the drive system is ensured by the control system [14]. The engine generates torque for forward and reverse travel, and can also act as a built-in battery generator, which, using the recuperation phenomenon, enables more efficient braking with simultaneous recuperation of the energy that is irretrievably lost in cars with combustion engines [15]. In an urban environment with frequent starting and stopping, this is of great importance, as it allows to minimize the energy losses associated therewith.

Due to the use of an electric motor only while driving, such cars convert the electrical energy stored in batteries.

The conversion of electrical energy into mechanical energy involves the emission of only small amounts of heat resulting from energy conversion losses, so that the operation of the drive in an electric car is characterized by a very high ecological cleanliness. Their overall CO_2 and pollutant emissions are limited to the electric power source emissions. Thus, unlike conventional and electric hybrid vehicles, they do not emit exhaust gases at the point of use, leading to the conversion of "end-of-pipe" emissions to those generated in power plants. In such a situation, total emissions from electric cars usually take place far away from cities, which has little impact on the health and well-being of the majority of the population, thus constituting a great advantage of such vehicles. These cars do not emit carbon monoxide or dioxide, nitrogen oxides or hydrocarbons, and in addition they significantly reduce noise levels.

The noise generated by electric vehicles in urban traffic reaches 40–60 dB, the same level as that of a functioning refrigerator.

For comparison, a diesel passenger car emits an average of 60–70 dB in urban traffic, a diesel bus – 90 dB, and a truck – 100 dB. The take-off jet emits 130 dB, i.e. the value assumed as the threshold of pain.

The paper [16] shows that electric cars generate much lower noise level than their combustion equivalents, especially in the speed range up to 50 km/h. Measurement of noise for two identical vehicles differing only in the power unit showed differences of 3 to 7 dB(A) depending on the speed of the vehicle, which means 2- to 5-fold reduction in noise level. For the quietest electric car used in the test (Nissan Leaf) and the noisiest combustion vehicle (BMW E30), the maximum difference was approx. 17 dB for speeds up to 50 km/h and 14 dB for speeds above 50 km/h. These two features of BEVs, i.e. zero emissions of harmful toxins and quiet operation, are particularly desirable in urban areas, because they somehow automatically solve the most burdensome effects of urban transport, except congestion. Unfortunately, the use of an electric car is associated with a number of inconveniences, the biggest of which are: the cost of purchase, poorly developed charging infrastructure and long charging time. Currently used battery technologies do not allow for immediate charging. Depending on the power source and battery capacity, charging time of discharged batteries can range from 20 min to more than 50 h (Table 1).

Table 1. Charging time of currently available cars with electric drive, depending on the type of charger [17].

Make and model	Total range (km)	Charging time [h]		
		Level 1	Level 2	DC charging
Mitsubishi i-Miev	100	12.4	3.1	0.29
smart fortwo electric drive	109	13.6	3.4	0.32
Ford Focus Electric	122	15.2	3.8	0.36
BMW-i3 BEV	130	16.2	4.05	0.38
Chevrolet Spark EV	132	16.4	4.1	0.39
Volkswagen e-Golf	134	16.6	4.15	0.39
Fiat 500e	135	16.8	4.2	0.4
Mercedes-Benz B250e	140	17.4	4.35	0.41
Kia Soul Electric	150	18.6	4.65	0.44
Nissan Leaf, 24 kW-h	135	16.8	4.2	0.4
Nissan Leaf, 30 kW-h	172	21.4	5.35	0.5
Tesla Model S, 70 kW-h	377	46.8	11.7	1.11
Tesla Model S, 85 kW-h	426	53	13.25	1.26
Tesla Model S, AWD 70D	386	48	12	1.14
AWD 85D	435	54	13.5	1.28
AWD 90D	435	54	13.5	1.28
AWD P85D	407	50.6	12.65	1.20
AWD P90D	407	50.6	12.65	1.20
Tesla Model X, AWD 90D	414	51.4	12.85	1.22
AWD P90D	402	50	12.5	1.19

Almost all consumer surveys have shown that one of the main obstacles to the implementation of e-mobility is the higher price of battery-powered electric cars compared to the analogous models with conventional combustion engines [18–21]. Table 2 compares the prices of selected BEVs with those of conventional cars with similar power and equipment.

Table 2. Price of selected new passenger cars with different engines in Poland (in PLN) [22].

Make	EV model	Price	Model with compression ignition	Price	Model with spark ignition	Price
Nissan	Leaf	128.000–165.000	Juke	72.000–84.000	Juke	72.000–89.000
Renault	Zoe	121.900–143.000	Clio	56.500–63.000	Twingo	46.000–62.000
Volkswagen	e-golf	162.890	golf	86.000–118.000	golf	80.000–94.000
Volkswagen	e-Up!	115.190	–	–	Up!	40.000

However, already today, from an economic point of view, the total cost of ownership (TCO = costs of purchasing a car (CAPEX) + costs of its use over a specific time (OPEX)) of an electric car is lower than that of a vehicle with a conventional engine [23]. However, as can be seen from [20, 24], consumers rarely make a purchasing decision based on a rational financial analysis. Most of them even believe that the costs associated with the operation of an electric car are much higher than in the case of vehicles with conventional propulsion [25]. Barriers related to the dissemination of battery cars also concern the infrastructure dedicated to this type of vehicles, i.e. public charging points. This problem is all the more serious because it is connected not only with a small number of such chargers, but also with limited access to them.

3.1 Application of Multi-criteria Point Method for the Assessment of BEV Electric Vehicles in the Urban Environment

In view of the issues presented above, an assessment of the use of electric vehicles in the urban environment was made. The analysis was made by comparing them with cars with different types of propulsion.

The compared vehicles are a single model with different energy sources and drive systems in the same market segment (B). They have the same or comparable total power, the same body type, the same type of drive (FWD) and gearbox (manual for conventional engines and automatic for electric motors).

The variants were:

variant 1 – spark-ignition engine,
variant 2 – compression-ignition engine,
variant 3 – electric motor,
variant 4 – Plug-In hybrid drive.

Table 3 presents technical parameters, emission performance, costs of, among others, purchase and other indices for the vehicles.

The multi-criteria analysis using the scoring method involves the aggregation of normalized values of features for each vehicle variant using predetermined scoring criteria and ordering the variants from the best score. This method facilitates the determination of the unambiguously best solution and clustering of solutions into categories depending on the score. It is also possible to score variants where scenarios or general objectives have been defined, e.g. technical, economic, environmental, for which several criteria have been identified.

Table 3. Criteria score value [26–30].

Costs	Volkswagen Golf, 5-door			
	Trendline 1.5 TSI ACT BlueMotion	Trendline 1.6 TDI	e-Golf	GTE 1.4 TSI PHEV
Curb weight [kg]	1.315	1.355	1.615	1.599
Load capacity [kg]	418–575	402–574	408–480	421–496
Total length [m]	4.36	4.26	4.27	4.27
Total width [m]	1.79	1.79	1.78	1.79
'Fuel' type	gasoline	diesel oil	electricity	gasoline/electricity
Average fuel consumption: gasoline [l]/diesel oil [l]/electrical energy [kWh] per 100 km	8.30	5.6	12.70	1.80
Maximum power [kW]	96	85	100	150
Maximum torque [NM]	200	250	290	350
Maximum speed [km/h]	210	198	150	222
Acceleration 0–100 km/h [s]	9.10	10.20	9.60	7.60
Total range [km]	820	1,087	231.00	883
Cost of purchase [PLN]*	82,960	88,360	165,690	120,550
Cost of 100 km [PLN]*	42.40	28.61	18.68	8.38
Number of dealers, service workshops**	1	1	0	0
Number of gas/charging stations (as of 02.2019)**	52	52	7	59
Time necessary to refill fuel or charge batteries (charging at alternating current at charging stations) [min]	2	2	320	147
Additional privileges such as the use of bus lanes, purchase subsidy, etc. [0–2]	0	0	2	1
CO_2 emission [g/km]	180.9	169.5	0	91.3
NOx+PMs emission [mg/km]	32.9	348.2	0	16.8
Noise emission at speeds up to 50 km/h [dB]	58.8	56.1	52.3	58.2
NGC [0–100]***	40	37	22	32

*Prices as of 11/02/2019
**Data for the city of Kielce
***Total emissions from the production of fuel, the vehicle and its use.
***NGC – an index encompassing CO, NO_x, HCs, PM10, SO_2, CO_2, CH_4, N_2O emissions during the production of fuel, the vehicle, its use and recycling/disposal.

The four variants (gasoline, diesel, hybrid, and electric car) were scored for environmental (objective 1), economic (objective 2), and technical (objective 3) criteria. The importance of the objectives c_k was determined in an expert discussion and the values are shown in Table 4. As assumed, the weights are between [0, 1] and their sum is 1.

Table 4. Objectives and criteria for the scoring and their weights.

Objective name (criteria group)	Weight (c_f)
Environmental	0.6
Economic	0.3
Technical	0.1

For each objective, appropriate partial scoring criteria were defined: three for the first one, two for the second one and five for the third one.

The scores were normalized to facilitate the comparison as per the procedure. Aggregated scores were determined by first multiplying weights of individual partial criteria by normalized scores for the criterion. Next, the scores within objectives for each solution were aggregated and multiplied by weights of individual objectives. Ultimately, the values for each solution were aggregated (Table 5).

Table 5. Synthetic score indices for the individual variants.

Objective	Variants			
	1	2	3	4
Objective 1	0.16	0.17	0.60	0.16
Objective 2	0.16	0.17	0.14	0.26
Objective 3	0.08	0.09	0.02	0.06
W(v)	0.40	0.42	0.76	0.49

Based on the analysis of the data in Table 5, variant 3 (electric car) had the largest index value. This is the best variant with the largest value of the index if all criteria (environmental, economic and technical) are considered together. It is worth mentioning at this point that the emission performance of cars that was taken into account in the calculations concerned only their use.

Using the same procedure as above and the data from Table 3, the confronted cars were evaluated for those cases where the emission performance of the cars included those produced during the production of fuel, manufacture of the vehicle and its use. In such a case, the fourth variant, a hybrid Plug-In type car had the highest index (Table 6).

Table 6. Synthetic score indices for individual variants - case 2.

Objective	Variants			
	1	2	3	4
Objective 1	0.42	0.34	0.57	0.45
Objective 2	0.16	0.17	0.14	0.26
Objective 3	0.08	0.09	0.02	0.06
W(v)	0.66	0.59	0.73	0.77

4 Conclusions

Nowadays, the automotive sector offers a wide range of cars with different technical parameters and energy sources used: gasoline, diesel oil, LPG, biofuels, electric current, etc. In this paper a multi-criteria scoring was used for the very first time to evaluate the operation of cars with different types of propulsion in urban environment. Analyzing the obtained results, it should be stated that electric cars are most advantageous, taking into account various factors, i.e. technical, economic and environmental, in use in the city nowadays. Taking into account only the emission performance of the vehicle at its place of use, the BEV type electric car was the best performer. When the emissions from the production of 'fuel', manufacture of the vehicle and its use were used in the calculations, the best variant was the Plug-In hybrid car. Still in this case, the use of purely electric cars in the city is a better alternative to combustion engine cars.

In conclusion, despite the difficulties associated with electric cars (purchase cost, small range, long battery charging time), these vehicles can be a key factor in building a sustainable mobility system and everything points to the fact that they will break Europe's longstanding dependence on internal combustion engines and petroleum fuels as means of meeting its transport needs. Increased use of electric vehicles, in particular when powered by renewable energy sources, can play an important role in achieving the EU's objective of reducing greenhouse gas emissions by 80–95% by 2050 and in the transition to a low-carbon economy in the future.

Under the current conditions, electric cars, through zero emissions of toxic compounds and CO_2 during their use, can contribute to improving air quality in urban agglomerations and improving the quality of life in those agglomerations.

References

1. 2018 Revision of World Urbanization Prospects, United Nations. Department of Economic and Social Affairs. https://www.un.org/development/desa/publications/2018-revision-of-world-urbanization-prospects.html
2. Together towards a competitive and resource efficient urban mobility, COM (2013). Report from the commission to the European Parliament, the Council, the European Economic and Social Committee and the Committee of the Regions
3. Ecology in car fleets. Ministry of the Environment, Energy Saving Trust - Leaseplan.pl

4. Ambient air pollution: A global assessment of exposure and burden of disease, WHO. https://www.who.int/phe/publications/air-pollution-global-assessment
5. Air quality in Europe – 2017 report, European Environment Agency. https://www.eea.europa.eu/publications/air-quality-in-europe-2017
6. European Commission press release of 16 November 2017. http://europa.eu/rapid/search.html
7. Signals 2016 - Towards clean and smart mobility, European Environment Agency. https://www.eea.europa.eu/publications/signals-2016
8. Directive of the European Parliament and of the Council on the limitation of emissions of certain pollutants into the air from medium-sized combustion plants SWD (2013) 532 final of 18/12/2013. https://eur-lex.europa.eu/
9. Outdoor air quality in urban areas, European Environment Agency. https://www.eea.europa.eu/airs/2018/environment-and-health/outdoor-air-quality-urban-areas
10. Data and statistics, World Health Organization. http://www.euro.who.int/en/health-topics/environment-and-ealth/ noise/data-and-statistics
11. Date and Maps, European Environment Agency. https://www.eea.europa.eu/data-and-maps
12. Report from the Commission to the European Parliament and the Council on the implementation of the Environmental Noise Directive in accordance with Article 11 of Directive 2002/49/EC. COM (2011) 321 final, European Commission, Brussels (2011)
13. Emissions of the main air pollutants in Europe, European Environment Agency. https://www.eea.europa.eu/data-and-maps/indicators/main-anthropogenic-air-pollutant-emissions/assessment-4
14. Nürnberg, M., Iwan, S.: Perspektywy stosowania samochodów elektrycznych w logistyce miejskiej na przykładzie realizacji usług kurierskich. http://www.ptzp.org.pl/files/konferencje/kzz/artyk_pdf_2018/T2/2018_t2_054.pdf
15. Electric vehicles in Europe, European Environment Agency. https://www.eea.europa.eu/publications/electric-vehicles-in-europe
16. Łebkowski, A.: Samochody elektryczne - dźwięk ciszy. Maszyny Elektryczne - Zeszyty Problemowe, 1 (2016)
17. Adderly, S.A., Manukian, D., Sullivan, T.D., Son, M.: Electric vehicles and natural disaster policy implications. Energy Policy **112**, 437–448 (2017)
18. Barisa, A., Rosa, M., Kisele, A.: Introducing electric mobility in Latvian municipalities: results of a survey. Energy Proc. **95**, 50–57 (2016)
19. Egbue, O., Long, S.: Barriers to widespread adoption of electric vehicles: an analysis of consumer attitudes and perceptions. Energy Policy **48**, 717–729 (2012)
20. Sovacool, B.K., Hirsh, R.F.: Beyond batteries: an examination of the benefits and barriers to plug-in hybrid electric vehicles (PHEVs) and a vehicle-to-grid (V2G) transition. Energy Policy **37**, 1095–1103 (2009)
21. Zhang, Y., Yu, Y., Zou, B.: Analyzing public awareness and acceptance of alternative fuel vehicles in China: the case of EV. Energy Policy **39**, 7015–7024 (2011)
22. INFO-EKSPERT Computer system - valuation of cars
23. Haddadian, G., Khodayar, M., Shahidehpour, M.: Accelerating the global adoption of electric vehicles: barriers and drivers. Electr. J. **28**(10), 53–68 (2015)
24. Turrentine, T.S., Kurani, K.S.: Car buyers and fuel economy? Energy Policy **35**, 1213–1223 (2007)
25. Krause, R.M., Carley, S.R., Lane, B.W., Graham, J.D.: Perception and reality: public knowledge of plug-in electric vehicles in 21 U.S. cities. Energy Policy **63**, 433–440 (2013)
26. Chargemap. https://chargemap.com/about/stats/poland
27. Next greencar. https://www.nextgreencar.com/emissions/ngc-rating/

28. Sendek-Matysiak, E.: Analysis of the electromobility performance in Poland and proposed incentives for its development. In: 2018 XI International Science-technical Conference Automotive Safety (2018). https://ieeexplore.ieee.org/document/8373338/
29. The Car Interior Noise Level Comparison site - Auto innengeräusch vergleich. http://www.auto-decibel-db.com/index_kmh.html
30. Volkswagen golf. https://www.vwgolf.pl/dane-techniczne/vw-golf-mk7/

Optimization as a Way to Improve the Efficiency of Transport Systems

MCDM as the Tool of Intelligent Decision Making in Transport. Case Study Analysis

Barbara Galińska[✉]

Lodz University of Technology, Lodz, Poland
barbara.galinska@p.lodz.pl

Abstract. Nowadays, the decision-making process is often affected by various risk factors, which may have negative consequences and lead to wrongful decisions. In these circumstances, the meaning of 'intelligence' aspect is gaining in importance as it highly enhances the possibility of making the right decision. One of the tool used in Intelligent Decision Making is Multiple Criteria Decision Making (MCDM) Methodology, application of which highly increases rationality in the selection process. It is composed of various methods, classified by the author as the intelligent tools. Additionally, Intelligent Decision Making models are very useful in various sectors of economy, including transportation sector. The typical decision problem may be e.g. the process of evaluating and selecting transportation systems and their components, such as logistics operators for example. Yet, the cooperation with the specialised, effective logistic operator may determine the success of the whole transportation process in the company. Therefore, the process of evaluating and selecting the main and the key logistics operators should be carefully considered and based on the intelligent approach. Pursuant to the main idea of the article, for Intelligent Decision Making can be used e.g. MCDM Methodology and its various methods. One of them – Promethee II will be applied, in order to make the right decision during the selection of the most desired variant – logistics operator.

Keywords: Decision making · Intelligence · Transportation · Multi-criteria decision making · Promethee II method · Logistics management · Management

1 Introduction

A natural human activity is decision making. Some of the decisions are made easily and have no further consequences, while others need to be made with the involvement of many research, which increase the rationality of selection. It mainly depends on the scope and the impact of such decision made in the long term. Thus, both in theory and practice of decision making, many various tools (methods, techniques, rules) may be applied and they all support decision-making process (selection of the most desired variant). One, is MCDM methodology, which enables rational selection. The author of this paper, classified it as a tool for Intelligent Decision Making, so it has been characterized in accordance with this idea, including the detailed description of its main methods.

This paper is the second article of the series dedicated to the issue of Intelligence and Intelligent Decision Making in logistics and transport. It was preceded by the article titled 'Logistics Megatrends and Their Influence on Supply Chains', which introduced the most important logistics megatrends and their influence on supply chains changes. The first article titled 'Intelligent Decision Making in Transport. Evaluation of Transportation Modes (Types of Vehicles) based on Multiple Criteria Methodology' indicated tools/techniques which may be applied to Intelligent Decision Making. Therefore, the empirical analysis of this paper presented selection of the most desired transportation modes (types of vehicles) for the trading company based on the Multiple Criteria Methodology for Intelligent Decision Making.

The paper elaborates on the issue of Intelligent Decision Making. Its main objective is to provide a thorough characteristics of MCDM methodology methods, classified as the tools for Intelligent Decision Making. Thereby, the objective of the empirical part of the paper is to evaluate the logistics operators for the company operating in the electronic industry, aiming at selection of the most desirable one. Such evaluation will be performed on the basis of the Promethee II method (applied to Intelligent Decision Making).

The paper is composed of 5 sections. The first one provides the short introduction as well as the aim of the paper. The second section includes the characteristics of MCDM as one of the tool for Intelligent Decision Making. Special attention has been paid to the description of the methods, which belong to this methodology, also being classified as the intelligent tools. Section 3 is focused on description of the decision situation, including the characteristics of logistics operators and their evaluation criteria. Section 4 includes the results of computational experiments generated with the applied method. In Sect. 5 the final conclusions are presented. The paper is supplemented by a list of references.

2 MCDM as One of the Tool for Intelligent Decision Making

2.1 Characteristics of Intelligent Decision Making

Decision Making is being defined as a cognitive process, which comprises a group of logically intertwined thinking and computational operations. They aim at solving the decision problem by selection of the most desired variant. Each decision process provides the final selection (sometimes referred to as a solution) [1].

In a lot of recognized models of decision making, such as presented by Simon [2, 3] or Żak [4], there is an 'Intelligence' expression used. Thus, to make a good decision, one has to decide from many alternatives. In order to increase the rationality of such selection, the process of Intelligent Decision Making should be involved, which is also a critical component of making good decisions. Above mentioned statement allows for presentation the issue of 'Intelligent Decision Making'. According to the author of this paper, it is being defined as a decision-making process, which involves both computers and their users. Simple Decision Making (SDM) become insufficient and needs to be replaced with the ones, based on the Artificial Intelligence (AI). Thereafter, Intelligent Decision Making, as the combination of AI and SDM, can be seen as a Decision Support System that provides access to data and knowledge-bases and/or conducts inference to support effective decision making in complex problem domains [1].

The main role of Intelligent Decision Making is adopting the best possible measures to the situation provided. Therefore, it should be characterized by certain intelligent behaviours, in particular [5–7]:

- learning from the experience, knowledge management,
- ability to implement the solutions based on the acquired knowledge,
- ability to understand ambiguity and contradiction,
- ability to response fast and appropriately in a new situation,
- ability to provide adequate reasoning in problems' solution in order to provide efficient decision making,
- ability to cope with troublesome situations,
- ability to apply and develop the knowledge, necessary to respond and react to the changes in the certain environment.

The research of Intelligent Decision Making has focused on decision-making component expanding to integration, from quantitative model to knowledge-based decision-making approach. It all substantially increases the value of Intelligent Decision Making models and encourages their implementation into economy practice. Intelligent models are becoming particularly useful in many economy sectors, including transportation sector, essential for each supply chain [8, 9].

In order to optimize and improve the quality of decision-making processes, different methods and techniques, classified as intelligent tools, can be applied [8]. One of them is Multiple Criteria Decision Making (MCDM), which is further described below.

2.2 Fundamentals of MCDM

One of the tools dedicated to Intelligent Decision Making is Multiple Criteria Decision Making (MCDM) also known as a Multicriteria Decision Aid (MCDA) or Multicriteria Decision Support. It is a field of study which derived from the operational research [10, 11], which also supports the decision-making process. MCDM starts with defining the objectives, creation of the variants and finally, selection of the most desired one. It provides thorough analysis of all the criteria defined by the Decision Maker (DM) in search for the most desired alternative [12].

Moreover, MCDM is focused on the development and implementation of decision support rules, methods, mathematical and programming tools in order to confront complex decision problems involving multiple criteria goals or objectives of conflicting nature (considering several – often contradictory – points of view) [13]. It has to be though emphasized, that MCDA techniques and methodologies, are not just some mathematical models aggregating criteria that enable one to make optimal decisions in an automatic manner. Instead, MCDM has a strong decision support focus. In this context the DM has an active role in the decision-modelling process, which is implemented interactively and iteratively until a satisfactory recommendation is obtained and fits the preferences and policy of a particular DM or a group of DMs [14].

MCDM is a methodology which is focused on assisting the DM in solving multiple criteria decision problems (MCDPs), i.e. situations in which having defined a set of actions (variants/solutions) A and a consistent family of criteria F, the DM tends to [13, 15, 16]:

- determine the subset of actions (variants, solutions) in A considered as the most desired in aspect of family of criteria F (choice problem),
- divide A into subsets according to concrete classification rules (sorting problem),
- rank actions (variants, solutions) in A from the best to the worst, according to F (ranking problem).

The set of A can be defined as set of objects, decisions, solutions, variants or actions, which are analysed and evaluated in decision process. It can be defined directly (in the form of a complete list) or indirectly (in the form of certain rules and formulas that determine feasible actions/variants/solutions, e.g. in the form of constraints). The consistent family of criteria F should be characterized by the following features [12, 15, 17]:

- it should provide a comprehensive and complete evaluation of all considered variants A,
- it should have a specific direction of DM's preferences – each criterion in A should have a specific direction of preferences (minimized or maximized) and should not be related with other criteria in F,
- the domain of each criterion in F should be disjoint with the domains of other criteria.

2.3 Main Streams in MCDM Research

The methodology of MCDM has a universal character and can be applied in various cases when a DM solves a MCDPs [18]. To solve the problems various methods can be used. These methods can be classified according to different criteria, including various aspects (Table 1).

There are particular methods of MCDM characterized below in accordance with the criterion: 'The manner of the preference aggregation'. In the author's opinion, these methods should be applied as intelligent tools for Intelligent Decision Making.

Table 1. Categories of MCDM methods (source: [21, 22]).

Classification criteria	Main streams (classification of methods and tools which are used to solve the multiple criteria decision problems)
The overall objective of the decision method correlated with the category of the decision problem	• multiple criteria choice methods (e.g. LBS, Topsis) • multiple criteria sorting methods (e.g. 4eMka) • multiple criteria ranking methods (e.g. Electre, AHP)
The moment of the definition of the DM's preferences	• methods with a'priori defined preferences (Electre, Promethee, UTA, Oreste) • methods with a'posteriori defined preferences (e.g. PSA) • interactive methods called multiple objective local evaluation methods based on the trial and error approach implemented in the specific interactions e.g. Light Beam Search – LBS [19], GDF, STEM)
The manner of the preference aggregation	• methods of American inspiration based on the utility function [20] referred to as the unique criterion of synthesis methods (e.g. UTA, AHP, ANP) • methods of the European/French origin, based on the outranking relation, also known as outranking synthesis methods, considering incomparability relation (e.g. Electre I–IV, Promethee, Oreste)
The form of the models, the model development process and their scope of application	• multiobjective mathematical programming • multiattribute utility/value theory • outranking relations • preference disaggregation analysis

2.4 Methods of American Inspiration, Based on the Utility Function

AHP (*Analytic Hierarchy Process*) Method

AHP is a method based on hierarchical analysis of decision problems. Using the method, it is possible to decompose a complex decision problem and make the final ranking for the definite set of variants. Pairwise comparisons, i.e. comparing elements in pairs with respect to a given criterion, are used for establishing priorities (or weights) among elements of the same hierarchical level. All the elements of the same hierarchical level are compared in pairs with respect to the corresponding elements in the next higher level, obtaining a matrix of pairwise comparisons. For representing the relative importance of one element over another, a scale for pairwise comparisons is introduced by Saaty [23–25], which defines and explains the values 1 through 9 assigned to judgments in comparing pairs of elements in each level with respect to a criterion in the next higher level.

AHP method facilitates the process of choosing the variants, which can be either some physical objects or some states that are represented by the defined variants. The method encompasses the multiple criteria approach which is based on the strategy of modeling the DM's preferences and on the assumption that the variants are comparable. Since the multiple criteria approach presumes that preferences are natural in the process of evaluation carried out by a person, the method takes into account the DM's preferences, which gives the evaluation a subjective character [26].

ANP (*Analytic Network Process*) Method
ANP method is being a development of AHP method. It does not require independence among elements as it forms a network structure to represent the problem, as well as pairwise comparisons to establish relations within the structure. The process of pairwise comparisons is performed in the same direction as in the AHP method [27].

UTA (*Utility Additive*) Method
UTA method belongs to the group of methods which adopt Multiple Attribute Utility Theory [28]. In UTA technique, the preference disaggregation approach refers to the analysis (disaggregation) of the global preferences (judgment policy) of the DM in order to identify the criteria aggregation model that underlies the preference result. UTA method refers to the philosophy of assessing a set of value or utility functions and adopts the preference disaggregation principle [29].

The DM needs to rank the best to the worst of the variants, by giving each variant a rank. Constraints can be imposed on the marginal utilities to respect (as much as possible) the given ranking. Properties such as transitivity will impose additional constraints [30].

2.5 Methods of the European/French Inspiration, Based on the Outranking Relation

ELECTRE (*Elimination and Choice Expressing the Reality*) Methods
The methods, which belong to the ELECTRE group, are well known and widely applied in multiple criteria decision making techniques. The choice of the final method selection is determined by the nature of the problem and the type of data concerned [31].

ELECTRE methods are the multiple criteria methods of ranking the finite set of variants which are evaluated with the application of the set of criteria [12, 13, 15, 17].

As the aim is to present the concept of the decision problem in the most realistic way, the indifference and preference thresholds are introduced to the selected method. The main rule in ELECTRE methods is comparison of each variant to the others, taking into account the thresholds which define the relation between these variants. Such way determines outranking relation of the one particular variant and enables incomparability with respect to the particular criterion when there is e.g. no enough information concerning preference relation.

PROMETHEE (*Preference Ranking Organization METHOD for Enriched Evaluation*) **Methods**

PROMETHEE method was introduced by Brans [32, 33] to preference rank a set of decision variants, based on their values over a number of different criteria. Put simply, a ranking of variants is based on the accumulative preference comparisons of pairs of variants' values over the different criteria [34].

In this method the DM needs to define all the criteria taken into account. Then all variants to be ranked, need to be evaluated according to those criteria. By specifying this preference information, the pairwise criterion preference degrees can be computed. From those preference degrees, unicriterion flows are computed. Finally, the criterion flows are aggregated into global flows by taking into account the relative importance of each criterion. All the method is based on the computation of preference degrees. A preference degree is a score (between 0 and 1) [30, 35], which expresses how an action is preferred over another action, from the DM's point of view. If the difference between the evaluations on a criterion is smaller than the indifference threshold, then no difference can be perceived by the DM between these two actions (i.e. the preference degree is 0). If the difference is higher than the preference threshold, then the preference is strong (i.e. the preference degree is 1). The preference function gives the value of the preference for differences between the indifference and preference threshold.

In the empirical part of the paper, the intelligent tool has been applied, which is Promethee II method, in order to make an intelligent decision concerning a selected variant – logistics operator – in the decision situation described.

3 Application of MCDM – Case Study

3.1 Major Features of the Considered Problem

The aim of the paper is to evaluate and select the most specialised logistics operator for the company operating in industry, namely Electronic Manufacturing Services (EMS). The company is acting under the contract agreement and it offers surface mount (leaded and lead-free) of electronic components. The company's ambition is to increase competitiveness of its business partners and products in offer. It belongs to the international Finnish company, possessing 10 branches located all over the world. The analysed company's branch is situated in central Poland – in town Sieradz, where over 800 workers are employed. The products in offer and services performed by the company are directed to domestic and foreign customers from automation, energy, IT and healthcare industries.

Customers' needs and requirements form the development activities of the company. Thus, in order to uphold high standards and further development, the company's authorities are aim to accelerate its business and to improve transportation services within the company.

In order to achieve so, the DM – Supply Chain Manager – in the decision situation described, made a decision about selecting the logistics operator (among current transport services providers), which would be competent enough, with the adequate financial and human resources and with all the necessary knowledge and experience to perform transport services in the most efficient way. Therefore, the DM has decided to perform a detailed evaluation of logistics operators, determined to the development of the cooperation with the most desired one (the most specialised logistics operator) and which would in the end improve the effectiveness and quality of the whole transportation process in the company.

3.2 Characteristics of the Considered Variants

The preliminary selection of logistics operators implemented ABC analysis, which grouped them according to quantitative criterion – number of transport assignments (number of shipments delivered by logistics operators in 2017). Out of eleven logistics operators, which are in constant cooperation with the company, five main logistics operators have been selected (Table 2).

Table 2. Variants – logistics operators – preliminary stage selection.

Variant – logistics operator	Number of shipments in 2017	Percentage	Cumulative percentage	Category
V1	852	25,52%	25,52%	A
V2	577	17,29%	42,81%	A
V3	564	16,90%	59,71%	A
V4	388	11,62%	71,33%	A
V5	190	5,69%	77,02%	A

All of the five selected operators have been taken into account in the second stage of the project, which is selection of the most specialised one. The problem of logistics operators selection has been defined as a multiple criteria ranking problem. The considered variants are logistics operators (selected in the preliminary stage) V1–V5 (Table 3).

Table 3. Variants – logistics operators – characteristics.

Variant	Characteristics
V1	Small Polish logistics company, located 7 km from the undertaking, 14 years on the market
V2	Large logistics company, part of German corporation. One of its headquarters is located within 20 km from the undertaking, existing 36 years on the market
V3	Large logistics company, part of American corporation. One of its headquarters is located within 25 km from the undertaking, existing 27 years on the market
V4	Large logistics company, part of Dutch corporation. One of its headquarters is located 13 km from the undertaking, existing 27 years on the market
V5	Large logistics company, part of German corporation. One of its headquarters is located within 18 km form undertaking, existing 14 years on the Polish market

3.3 Description of Evaluation Criteria

Until now, the company selected logistics operators only by considering price criterion – DM has never performed any detailed evaluation. It is against the idea of 'Intelligent Decision Making' which is based on complex analysis of the considered variants, including application of the various intelligent tools and additionally based on evaluation's criteria.

In this paper, the evaluation's criteria have been constructed according to the model proposed by Galińska and Żak [36], which distinguishes seven main evaluation criteria. The importance wages of the criteria were formulated on the basis of the interview with the DM, his preferences and aspirations (total amount of criteria weights has to be 1). Due to research limitations, criteria K1–K7 and their sub-criteria have been only enumerated, without any detailed descriptions (Table 4).

Table 4. Evaluation criteria in the case study.

Criterion	Sub-criterion	Weight of criterion
K1: Transportation Costs	The criterion is defined as the total/overall costs of moving goods from their origin to final destination, expressed in monetary units [PLN]	0.20
K2: Delivery Time	The criterion is defined as the total duration time of the delivery process, expressed in days	0.18
K3: Timeliness of Delivery	The criterion defines the share (percentage - wise) of on-time deliveries	0.15
K4: Reliability of a Transportation System	The criterion measures the share (percentage - wise) of damaged shipments in the delivery process	0.10
K5: Flexibility of a Delivery System	K5.1 – frequency of delivery (defined as a number of potential deliveries in a monthly period)	0.10
	K5.2 – minimal size of the shipment (expressed on the 1 to 10 point scale)	0
	K5.3 – capacity availability (average share of container's/trailer's capacity evaluable)	0
K6: Safety of a Delivery System	K6.1 – number of handling operations	0.08
	K6.2 – total period of the storage in the transhipment terminals	0
K7: Customer's Comfort	K7.1 – ability to monitor the shipments and its status information (expressed in point)	0.12
	K7.2 – additional services, including complaint processing	0.03
	K7.3 – payment periods (expressed in days)	0.04

Considering the nature of the case study described, some of the evaluation criteria have been found to be invalid (0 weight). Thus, they have been ignored in the further part of the computational experiments.

4 Computational Experiments – Ranking of the Variants with the Application of Promethee II Method

The ranking of the variants – logistics operators – has been performed with the application of the MAMCA Software, which is the implementation of Promethee II method. In accordance with the algorithm of the applied method, the evaluation matrix of each variants (V1, V2, …V5) has been constructed (Table 5). Then the DM's preference model has been defined in the process of naming the wages of criteria and thresholds: indifference and preference. The thresholds define the sensitivity of the DM to the changes of the criteria values and the weight expresses the importance of each criterion. The matrix includes evaluations of all variants on all criteria and sub-criteria.

Table 5. The Evaluation Matrix in the case study.

Criteria		V1	V2	V3	V4	V5
K1 [PLN]		2700	1550	1370	2300	889
K2 [DAYS]		1	1	3	3	5
K3 [%]		1	0.95	0.95	0.90	0.95
K4 [%]		0	0	0.10	0.10	0.05
K5	K5.1 [POINTS]	16	20	12	12	20
K6	K6.1 [POINTS]	2	3	4	4	2
K7	K7.1 [POINTS]	5	5	4	4	5
	K7.2 [POINTS]	2	5	3	4	5
	K7.3 [DAYS]	60	60	30	21	21

Taking into account that each variant is compared to the other variants, two outranking flows are defined: the positive outranking flow and the negative outranking flow. The positive outranking flow shows how a variant outranks all others. It shows its relevance character: the higher it is, the better the variant. The negative outranking flow shows how a variant is outranked by others. It shows its weakness character: the lower it is, the better the variant. The Promethee II method eliminates all incomparabilities by the evaluation of the net outranking flow. The ranking of the variants is given by the net outranking flow, where the variant with the highest value is considered as the best variant by the evaluation process. The rest of the variants are ranked accordingly, from the highest to the lowest net outranking flow. The net outranking flow is the balance between the positive and the negative outranking flows. The higher the net flow, the better the variant.

Computational experiments led to indicate the best (most specialized) logistics operator in the case study described. The final ranking identifies the ranking of particular variants relative to each other (Fig. 1).

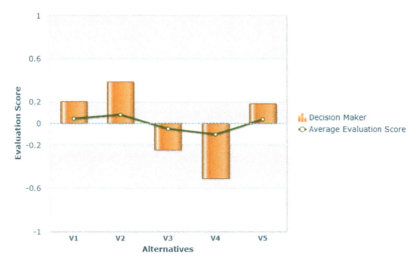

Fig. 1. Final ranking of variants

The final ranking obtained in the process of computational experiments indicated the most and the least desired logistics operators in the case study described.

Analysing all the final results, the most specialised variant is V2. It is a large logistics company, which belongs to the German corporation. This logistics operator is featured by such characteristics, as: average transportation costs, low delivery time, average timeliness of delivery, high flexibility of a transportation system and very high customer's comfort. The author also advises that for the cheaper deliveries (lower transportation costs) variant V5 should be selected.

5 Final Conclusions

The paper presents a detailed description of MCDM Methodology, which is regarded by the author as a tool for Intelligent Decision Making.

As the article is the second of the series dedicated to the issue of 'Intelligence' and 'Intelligent Decision Making' in logistics and transport, it contains a characteristics of Intelligent Decision Making features. Further, the author provides description of MCDM Methodology and its particular methods. The author establishes various criteria of the classification, on the basis of: 'The manner of the preference aggregation' criterion, in order to characterise the methods based on the utility function (American approach) and on the outranking relation (European approach). The empirical part of the paper involves the evaluation and selection process of the most specialised logistics operator for the company operating in electronic industry Electronic Manufacturing Services.

The author, with the reference to her previous works assumed that evaluation of logistics operators should have a multiple criteria character and thus, she applied the principles of Multiple Criteria Decision Making, in particular the Promethee II method.

In this approach the author attempts to use the tools and methods of Multiple Criteria Decision Making, recognized as a part of the Artificial Intelligence, which is consistent with the main idea of the paper.

In practical terms the author argues that the best (the most specialised) logistics operator is variant V2, recognized as being the most effective. Thus, it is recommended to select logistics operator V2 as the 'main supplier' for the DM.

In the author's opinion further research should be carried out in two directions:

1. Further analysis of intelligent tools/methods applied to Intelligent Decision Making.
2. Carrying out a comparison between intelligent and classical approach, including their tools used in decision-making process.

References

1. Lu, J., Zhang, G., Ruan, D., Wu, F.: Multi-Objective Group Decision Making: Methods, Software and Applications with Fuzzy Set Techniques. Imperial College Press, London (2014)
2. Simon, H.A.: The New Science of Management Decision. Prentice-Hall, Englewood Cliffs (1977)
3. Simon, H.A.: Administrative Behavior. The Free Press, New York (1997)
4. Żak, J.: The concept of intelligent decision making in logistics. In: CLC 2012 Conference, Jeseník, Czech Republic (2012)
5. Turban, E., Aronson, J.E., Liang, T.P.: Decision Support Systems and Intelligent Systems. Pearson Prentice Hall, New Jersey (2005)
6. Sierpiński, G., Staniek, M., Celiński, I.: New methods for pro-ecological travel behavior learning. In: 8th International Conference of Education, Research and Innovation (ICERI 2015), Seville, Spain, pp. 6926–6933 (2015)
7. Sierpinski, G., Staniek, M.: Education by access to visual information methodology of moulding behaviour based on international research project experiences. In: 9th International Conference of Education, Research and Innovation (ICERI 2016), Seville, Spain, pp. 6724–6729 (2016)
8. Galińska, B.: Intelligent decision making in transport. Evaluation of transportation modes (types of vehicles) based on multiple criteria methodology. In: Sierpiński, G. (ed.) Advances in Intelligent Systems and Computing: Integration as Solution for Advanced Smart Urban Transport Systems, vol. 844, pp. 161–172. Springer, Cham (2019)
9. Sierpiński, G., Celiński, I., Staniek, M.: The model of modal split organisation in wide urban areas using contemporary telematic systems. In: 3rd International Conference on Transportation Information and Safety, Wuhan, China, pp. 277–283 (2015)
10. Hillier, F., Lieberman, G.: Introduction to Operations Research. McGraw-Hill, New York (1990)
11. Żak, J.: Application of operations research techniques to the redesign of the distribution systems. In: Dangelmaier, W., Blecken, A., Delius, R., Klöpfer, S. (eds.) Advanced Manufacturing and Sustainable Logistics. Paderborn, Germany (2010)
12. Roy, B.: Decision-aid and decision making. Eur. J. Oper. Res. **45**, 324–331 (1990)
13. Figueira, J., Greco, S., Ehrgott, M.: Multiple Criteria Decision Analysis. State of the Art Surveys. Springer, New York (2005)

14. Doumpos, M., Evangelos, G.: Multicriteria Decision Aid and Artificial Intelligence: Links, Theory and Applications. Wiley, Hoboken (2013)
15. Vincke, P.: Multicriteria Decision-Aid. Wiley, Chichester (1992)
16. Żak, J.: Multiple Criteria Decision Aiding in Road Transportation. Poznan University of Technology Publishers, Poznan (2005)
17. Roy, B.: The outranking approach and the foundations of ELECTRE methods. In: Bana e Costa, C. (ed.) Readings in Multiple Criteria Decision Aid. Springer, Berlin (1990)
18. Żak, J.: Comparative analysis of multiple criteria evaluations of suppliers in different industries. Transp. Res. Procedia **10**, 809–819 (2015)
19. Książek, M., Nowak, P., Kivrak, S., Rosłon, J., Ustinovichius, L.: Computer-aided decision-making in construction project development. J. Civ. Eng. Manag. **21**(2), 248–259 (2015)
20. Keeney, R., Raiffa, H.: Decisions with Multiple Objectives. Preferences and Value Tradeoffs. Cambridge University Press, Cambridge (1993)
21. Żak, J.: Design and evaluation of transportation systems. In: Sierpiński, G. (ed.) Advances in Intelligent Systems and Computing: Advanced Solutions of Transport Systems for Growing Mobility, vol. 631, pp. 3–29. Springer, Cham (2018)
22. Pardalos, P.M., Siskos, Y., Zopounidis, C.: Advances in Multicriteria Analysis. Kluwer Academic Publishers, Dordrecht (1995)
23. Saaty, T.: How to make a decision: the analytic hierarchy process. Eur. J. Oper. Res. **48**, 9–26 (1990)
24. Saaty, T.: The Analytic Hierarchy Process: Planning, Priority Setting. Resource Allocation. McGraw-Hill, New York (1980)
25. Saaty, T.: Transport planning with multiple criteria: the analytic hierarchy process applications and progress review. J. Adv. Transp. **29**, 81–126 (1995)
26. Downarowicz, O., Krause, J., Sikorski, M., Stachowski, W.: Zastosowanie metody AHP do oceny i sterowania poziomem bezpieczeństwa złożonego obiektu technicznego. Wydawnictwo Naukowe Politechniki Gdańskiej, Gdańsk (2000)
27. Saaty, T.: Decision Making with Dependence and Feedback. The Analytic Network Process. RWS Publications, Pittsburgh (1996)
28. Jacquet-Lagreze, E., Siskos, J.: Assessing a set of additive utility functions for multi criteria decision-making, the UTA method. Eur. J. Oper. Res. **10**, 151–164 (1982)
29. Żak, J., Bieńczak, M., Fierek, S., Kruszński, M., Ratajczak, J., Sawicka, H., Żmuda-Trzebiatowski, P.: Zastosowanie metodyki wielokryterialnego wspomagania decyzji (WWD) do rozwiązywania wybranych problemów decyzyjnych związanych z zarządzaniem miastem. Wydawnictwo Naukowe Politechniki Poznańskiej, Poznań (2010)
30. Ishizaka, A., Nemery, P.: Multi-Criteria Decision Analysis: Methods and Software. Wiley, New York (2013)
31. Nowak, M.: Preference and veto thresholds in multicriteria analysis based on stochastic dominance. Eur. J. Oper. Res. **158**, 339–350 (2004)
32. Brans, J.P., Mareschal, B., Vincke, P.H.: PROMETHEE: a new family of outranking methods in MCDM. In: Brans, J.P. (ed.) International Federation of Operational Research Studies, IFORS 1984, North Holland, Amsterdam, pp. 470–490 (1984)
33. Brans, J.P., Vincke, P.H., Mareschal, B.: How to select and how to rank projects: the PROMETHEE method. Eur. J. Oper. Res. **24**, 228–238 (1986)
34. Kun-Huang, H.: Quantitative Modelling in Marketing and Management. World Scientific Publishing, Singapore (2014)
35. Trzaskalik, T.: Wielokryterialne wspomaganie decyzji. Przegląd metod i zastosowań. Zeszyty Naukowe Politechniki Śląskiej, seria Organizacja i Zarządzanie **74**, 239–263 (2014)
36. Żak, J., Galińska, B.: Design and evaluation of global freight transportation solutions (Corridors) analysis of a real world case study. Transp. Res. Proc. **30**, 350–362 (2018)

AHP as a Method Supporting the Decision-Making Process in the Choice of Road Building Technology

Izabela Skrzypczak[1(✉)], Wanda Kokoszka[1], Tomasz Pytlowany[2], and Wojciech Radwański[2]

[1] Rzeszów University of Technology, Rzeszów, Poland
izas@prz.edu.pl
[2] Krosno State College, Polytechnic Institute, Krosno, Poland
tompyt@pwsz.krosno.pl

Abstract. The article presents a choice of road building technology using the AHP method and the Hellwig method. Based on the analyses carried out with the use of the Saaty and Hellwig methods it was found that the achieved values of ranking for asphalt and concrete pavements are close to each other, which may indicate that both technologies are comparable within the criteria taken for analyses in relation to accepted technological-technical and usability features. When choosing the road building technology one shouldn't be limited to building costs alone. It's necessary to consider the costs of 30–40 year exploitation and maintenance of the road. The main task when choosing the road building technology is to build the road of a quality that will enable many years of its exploitation. Based on the analyses carried out with the use of the Hellwig and AHP methods it can be assumed that both technologies. the asphalt one and the concrete one can be competitive because it leads to the progress and development of both of them. Considering the increasing growth of traffic, concrete may be perceived as a technical and economic alternative in comparison to asphalt pavement; which is confirmed in the presented analyses.

Keywords: Pavement · Hellwig · Saaty · AHP

1 Introduction

Pavement engineering is the selection of design, materials, and construction practices to ensure satisfactory performance over the projected life of the pavement. Pavement users are sensitive to the functional performance of pavements - smoothness and skid resistance - rather than structural performance. Pavements, as a general rule, develop distresses gradually over time under traffic loading and environmental effects. An exception is when poor material choices or construction practices cause defects before or shortly after the pavement is put into service.

State highway agencies generally select pavement type either by economy, service life, life cycle cost or all three. Service lives for concrete pavements of so much as 20 to 40 years, or generally 1,5–2 times the service life than asphalt pavements designed and built to similar standards. In general, concrete is selected for heavily trafficked pavements and typically carries four times as much daily truck traffic as asphalt pavements. Thus, it should come as no surprise that rigid pavements are often used in urban, high traffic areas. But, naturally, there are trade-offs. For example, when a flexible pavement requires major rehabilitation, the options are generally less expensive and quicker to perform than for rigid pavements. Respective life cycle costs of pavement designs depend greatly on material costs at the time of construction, but concrete pavements often have significantly lower maintenance costs although initial construction costs higher. The both technologies for road construction, e.g. an asphalt one and a concrete one, that exist in the market of road infrastructure at the moment have their advantages and disadvantages [1–21].

When choosing the road building technology one shouldn't be limited to building costs alone. It's necessary to consider the costs of 30–40 year exploitation and maintenance of the road. The main task when choosing the road building technology is to build the road of a quality that will enable many years of its exploitation [5–13]. Based on the analyses carried out with the use of the Saaty [22–26] and Hellwig [27, 28] methods it can be assumed that both technologies. the asphalt one and the concrete one can be competitive because it leads to the progress and development of both of them. Considering the increasing growth of traffic, concrete may be perceived as a technical and economic alternative in comparison to asphalt pavement; which is confirmed in the presented analyses.

Concrete overlays of various types have an important part to pave in the upgrading and maintaining of the overall network. Accelerated on fast track paving materials and techniques have been developed to get traffic into pavements more quickly.

Properly designed and built out pavements can provide some years of service with little or no maintenance. Concrete pavement has a higher initial cost than asphalt but lasts longer and has lower maintenance costs [1, 21].

In some cases, however, design or construction errors or poorly selected materials have considerably reduced pavement life. It is therefore important for pavement engineers to understand materials selection, mixture proportioning, design and detailing, drainage, construction techniques, monitoring and pavement performance [1–21, 30]. It is also important to understand the theoretical framework underlying commonly used design procedures, and to know the limits of applicability and the mechanical properties of this both pavements [29]:

- Flexible pavement from rigid pavement in terms of load distribution. In flexible pavements load distribution is primarily based on layered system. While, in case of rigid pavements most of the load carries by slab itself and slight load goes to the underlying strata,
- Structural capacity of flexible pavement depends on the characteristics of every single layer. While, the structural capacity of rigid pavements is only dependent on the characteristics of concrete slab,
- In flexible pavements, load intensity decreases with the increase in depth. Because of the spreading of loading in each single layer. While, in case of rigid pavement maximum intensity of load carries by concrete slab itself, because of the weak underlying layer,

- In flexible pavement deflection basin is very deep, because of its dependency on the underlying layers,
- Flexible pavement has very low modulus of elasticity (less strength). Modulus of elasticity of rigid pavement is very high, because of high strength concrete and more load bearing capacity of the pavement itself. Than compared to flexible pavements,
- In flexible pavements, underlying layers play very important role. Therefore, more role are playing underlying layers. Maximum role is playing by the top layer (that is slab) by itself, in case of rigid pavements, slight function they have underlying layers.

Therefore, the aim of the work will be to present the possibilities of using a multi-criteria decision-making method of Analytical Hierarchy Process (AHP) in analysis of comparison between flexible and rigid pavement properties.

2 AHP Method

The AHP method is a commonly used tool for undertaking complex decisions based on a considerable number of criterions. This process includes comparative analyses of flexible and rigid pavements. The basis for the AHP method is hierarchic decomposition of the criterions of evaluation.

The structure of hierarchy of importance in the AHP method is determined in advance. The first step is formulating the process of decision making, then evaluation criteria and solution options are set up based of the expert's knowledge.

While using the AHP method the importance for evaluated parameters and features will be determined and results obtained will enable creating the comparative ranking of pavements in accordance with Hellwig method.

The AHP method takes into consideration specifics of psychological processes of evaluation which, first of all are of the relation and hierarchic character.

Numerous uses of this method for supporting economic, technical or social decisions confirm its usability, especially in cases where considerable evaluation criterions are of qualitative character and experience of the evaluator are the main source of evaluations of the objective character.

In its general form the AHP is a nonlinear framework for carrying out both deductive and inductive thinking without use of the syllogism by taking several factors into consideration simultaneously and allowing for dependence and for feedback, and making numerical tradeoffs to arrive at a synthesis or conclusion.

Aggregation of the evaluations in AHP method is done in accordance with additive usability function, which synthesizes participation of criterions and value of the degree of fulfillment of the fractional function of the purpose by each of these criterions. Evaluations of the degree of fulfillment of these criterions for the decision options under consideration are obtained by the binary comparison. The binary combination is based on the scale proposed by Satty [22–24] for element comparison in Fig. 1.

The pairwise comparison is the fundamental component of the AHP process. For each pairing within each criterion, the better option is awarded a score, again, on a scale between 1 (equally good) and 9 (absolutely better), whilst the other option in the pairing is assigned a rating equal to the reciprocal of this value (Table 1).

Table 1. Saaty scale for various elements comparison Saaty [22–24].

Scale	Judgment of preference	Description
1	Equally important	Two factors contribute equally to the objective
3	Moderately important	Experience and judgment slightly favour one over the other
5	Important	Experience and judgment strongly important favour one over the other
7	Very strongly important	Experience and judgment strongly important favour one over the other
9	Extremely important	The evidence favouring one over the other is of the highest possible validity
2, 4, 6, 8	Intermediate preference between adjacent scales	When compromised is needed

Each score records how well option "X" meets criterion "Y". Afterwards, the ratings are normalized and averaged. Twenty fife (25) experts provide their judgment of the relative importance of one indicator against another. Their results were normalized and examined with the Consistency Ratio test (CR). When CR value is less than 0.20, consistency of the comparison is appropriate [24]. Some of the authors accept 0.10 for CR upper limit.

The AHP method consists of following steps [24]:

- determine the problem,
- define the decision hierarchy from top to bottom considering the purpose of the decision.
- define the pair-wise comparison matrix (1), (2), (3):

Matrix $M = (m_{ij})$ is said to be consistent if $m_{ij} \cdot m_{ik} = m_{ik}$ and its principal eigenvalue (λ_{max}) is equal to n. The general eigenvalue formulation is (formulas 1 to 3):

$$M \cdot w = \begin{vmatrix} \frac{w_1}{w_1}=1 & \cdots & \frac{w_1}{w_{n-1}} & \frac{w_1}{w_n} \\ \frac{w_{n-1}}{w_1} & \frac{w_{n-1}}{w_{n-1}}=1 & \cdots & \frac{w_{n-1}}{w_n} \\ \vdots & & & \vdots \\ \frac{w_1}{w_n} & \frac{w_{n-1}}{w_n} & & \frac{w_n}{w_n}=1 \end{vmatrix} \begin{vmatrix} w_1 \\ \cdot \\ \cdot \\ \cdot \\ w_{n-1} \\ w_n \end{vmatrix} \qquad (1)$$

and $i, j = 1, 2, 3, \ldots n$

$$M \cdot w = n \cdot \begin{vmatrix} w_1 \\ w_2 \\ w_3 \\ \vdots \\ w_{n-2} \\ w_{n-1} \\ w_n \end{vmatrix} = \lambda_{max} \cdot w \qquad (2)$$

$$\lambda_{max} = \frac{1}{n} \cdot \sum_{i=1}^{n} \frac{(M \cdot w)_i}{w_i} \qquad (3)$$

- Apply consistency test (formula 4).

  For measure consistency index (CI) adopt the value (4):

$$CI = \frac{\lambda_{max} - n}{n - 1} \qquad (4)$$

The CR is obtained by comparing the CI with an average random consistency index RI (formula (5) and value for RI from Table 2).

$$CR = \frac{CI}{RI} \qquad (5)$$

RI is the random index (Table 2).

Table 2. Random index matrix of the same dimension [22].

Number of criteria	2	3	4	5	6	7	8	9	10
RI	0.00	0.58	0.90	1.12	1.24	1.32	1.41	1.45	1.49

Number of criteria = number of parameters compared

- Calculate relative local and global weights of each main and sub-factors.
- Synthesis of priorities - for this step obtain the principal right eigenvector and largest eigenvalue.

3 Materials and Methods

Two road building technologies exist in the road infrastructure market: an asphalt one and a concrete one, and they are both has its advantages and disadvantages [1–21], but which technology to choose, which one is better? The article presents the comparative analyses of this two road building technologies, an asphalt one and a concrete one. For the calculations, the AHP method was used [22–26] and Hellwig method [27, 28]. The Hellwig method is discussed in more detail in the article [27]. For this analysis, in accordance with Fig. 1, weights accordance with AHP method for the criterions and indicators important for the choice of road building technology were determined.

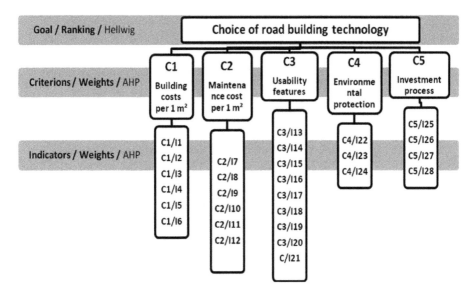

Fig. 1. Hierarchy structured decision

In order to develop the rankings of the choice of road building technology in the area of research 28 indicators describing the properties of these both road building methods was designated. These factors represent the five criterions of issues and served as diagnostic indicators used for further research. All elements under each criterions were set based on literature and expert knowledge [1–21, 29]. A list of the 28 indicators charactering the road building technology is shown in Table 6.

By using analytic hierarchy process, the preference of all data's corresponding to each criterion can be evaluated and given final ranking of pavements. The questionnaire conducted between the dates March 2018–July 2018 is answered by 25 experts. Data were collected from the experts in their offices and via email. They are asked to compare the criteria at a given level on a pair-wise basis to identify their relative precedence. The back ground concerning with criterions of experts outlined in Table 4.

The calculations were made using an MS Excel spreadsheet. Based on Saaty scale, the different weight has been attributed to the criterions and for indicators. The importance of the criterions (calculation weights) are given in Tables 3, 4 and 5.

In this study, weight were assigned to the different thematic indicators classes and layers based on their relative influence and contribution to the choice of road building technologies. The overlay technique was employed to the criteria and indicators to determine the weights. All processes were done AHP method for to the comparative analyses of road building technologies.

The described algorithm for calculating weights for individual criteria was repeated also for all indicators for the particular criterions (Table 6).

Table 3. Criteria matrix - (M).

	C1	C2	C3	C4
C1	1.000	9.000	1.000	1.000
C2	0.111	1.000	1.000	1.000
C3	1.000	1.000	1.000	1.000
C4	1.000	1.000	1.000	1.000
C5	1.000	1.000	1.000	1.000
Sum	4.111	13.000	5.000	5.000

Table 4. Normalization of criteria matrix and values of weights - (w) for criteria's.

	C1	C2	C3	C4	Weights – (w)
C1	0.243	0.692	0.200	0.200	0.307
C2	0.027	0.077	0.200	0.200	0.141
C3	0.243	0.077	0.200	0.200	0.184
C4	0.243	0.077	0.200	0.200	0.184
C5	0.243	0.077	0.200	0.200	0.184
Sum	1.000	1.000	1.000	1.000	1.000

Table 5. The general eigenvalues and Consistency Ratio test for analyzed criteria.

Products of matrix = $M \cdot w$	$\frac{(M \cdot w)_i}{w_i}$
2.126	6.923
2.126	15.102
2.126	11.554
2.126	11.554
2.126	11,554
Consistency Ratio test CR	0.168

Past experience and the back-ground of the experts are utilized in the determination of the criteria and 28 important indicators to be used for destination selection are established. The main 5 criteria's were used in evaluation and decision hierarchy is established accordingly. Decision hierarchy structured are provided in Fig. 1. There are two levels in the decision hierarchy structured for destination ranking problem.

The overall goal of the decision process determined as determining the ranking of choice of building roads technology is in the first level of the hierarchy. The criteria are on the second level and the indicators are on the third level of the hierarchy. After forming the decision hierarchy for the problem, the weights of the criteria to be used in evaluation process are calculated by using AHP method. In this phase, the experts in the expert team are given the task of forming individual pairwise comparison matrix by using the Saaty's 1–9 scale by Table 1.

Mean of experts choice values are calculated to obtain the pairwise comparison matrix on which there is a consensus (Table 6). The calculation of the ranking on choice of building roads technology is made by using the results obtained from the computations based on the pairwise comparison matrix and by Hellwig method and are presented in Table 6.

Table 6. Calculating weights w and CR - values of indicators obtained with AHP method.

No.	Indicator	Unit	Bituminous pavement	Concrete pavement	Weight	CR
C1	Building costs per 1 m^2					
C1/I1	KR1	PLN/m2	169.17	177.39	0.167	−0.355
C1/I2	KR2	PLN/m2	204.54	189.79	0.167	
C1/I3	KR3	PLN/m2	249.14	266.84	0.167	
C1/I4	KR4	PLN/m2	288.92	278.69	0.167	
C1/I5	KR5	PLN/m2	319.27	302.39	0.167	
C1/I6	KR6	PLN/m2	346.62	316.96	0.167	
C2	Maintenance costs per 1 m^2					
C2/I7	KR1	PLN/m2	533.28	370.00	0.167	−0.355
C2/I8	KR2	PLN/m2	653.10	396.91	0.167	
C2/I9	KR3	PLN/m2	707.70	481.41	0.167	
C2/I10	KR4	PLN/m2	764.12	516.31	0.167	
C2/I11	KR5	PLN/m2	810.88	570.55	0.167	
C2/I12	KR6	PLN/m2	854.54	605.46	0.167	
C3	Usable features					
C3/I13	Longitudinal evenness	–	1	0.6	0.197	−0.192
C3/I14	Furrowing	–	0.1	1	0.189	
C3/I15	Anti-slip properties/Motorways	Conclusive factor of friction	0.39	0.51	0.097	
C3/I16	Anti-slip properties/State roads	Conclusive factor of friction	0.44	0.47	0.097	
C3I17	Noise [21]	[1/Average form (C3I17-1 to C3I17-6)] * 100	10.20	10.23	0.055	
C3I17-1	Motorways/50 km/h	Index CPX	92.8	90.1		
C3I17-2	Motorways/80 km/h	Index CPX	100.1	97.4		
C3I17-3	Motorways/110 km/h	Index CPX	104.6	102.4		
C3I17-4	State roads/50 km/h	Index CPX	90.4	92.1		
C3I17-5	State roads/80 km/h	Index CPX	97.8	100		
C3I17-6	State roads/110 km/h	Index CPX	102.5	104.6		
C3I18	Colour of pavement/Visibility		0.7	1	0.022	
C3/I19	Colour of pavement/Surface heating	Degrees C	46.97	36.08	0.023	
C3/I20	Resistance to permanent deformation	–	0.7	1	0.189	
C3/I21	Breaking distance at 100 km/h	[1/average (C3/I21-1to C3/I21-2)] * 100	1.20	1.38	0.133	
C3/I21-1	Wet surface	m	109	96		
C3/I21-2	Dry surface	m	58	49		

(*continued*)

Table 6. (*continued*)

No.	Indicator	Unit	Bituminous pavement	Concrete pavement	Weight	CR
C4	*Environmental protection*					
C4/I22	Emission of CO2	(1/SUMA (C4/I22-1 to C4/I22-4) * 10000)	11.3	2.56	0.3270	−0,123
C4/I22-1	Emission of CO2 From asphalt and concrete production (kg of CO2/ton) [17]	kg CO2/t	27.4	694		
C4/I22-2	Emission of CO2 from production of 1 t of mineral-asphalt mixture and 1 t of concrete (kg of CO2/t) [18]	kg CO2/t	10.3	107.3		
C4/I22-3	Emission of CO2 from building 1 km of asphalt and concrete motorway (kg of CO2/km) [20]	kg CO2/km	347	1497		
C4/I22-4	Emission of CO2 from maintenance of 1 km of asphalt and concrete motorway (kg of CO2/km) [20]	kg CO2/km	500	1610		
C4/I23	Index of influence of building1 km of motorway on the environment	1/SUMA (C4/I23-1to C4/I23-5) * 10000000	5.81	3.61	0.3290	
C4/I23-1	Greenhouse effect potential (GWP) [19]	[kg of CO2 equivalent]	1712501.5	2765765		
C4/I23-2	Stratospheric ozone layer deterioration potential (ODP) [19]	[kg of CFC-11 equivalent]	0.395	0.13		
C4/I23-3	Photo oxidant synthesis potential (POCP) [19]	[kg of C2H4 equivalent]	422	384.5		
C4/I23-4	Acidification – potential (AP) [19]	[kg of SO2 equivalent]	8353.5	6426		
C4/I23-5	Eutrophication – potential (EP) [19]	[kg PO3-4]	1248	1092		
C4/I24	Index of influence on the repair and exploitation of 1 km of motorway on the environment	1/SUMA (C5/I24-1 to C5/I24-5) * 1000000	5.81	3.61	0.3440	
C4/I24-1	Greenhouse effect potential (GWP) [19]	[kg of CO2 equivalent]	996135	62245.5		
C4/I24-2	Stratospheric ozone layer deterioration potential (ODP) [19]	[kg of CFC-11 equivalent]	0.225	0.01		
C4/I24-3	Photo oxidant synthesis potential (POCP)	[kg of C2H4 equivalent]	294	46		
C4/I24-4	Acidification – potential (AP) [19]	[kg of SO2 equivalent]	5638.5	267.5		
C4/I24-5	Eutrophication – potential (EP) [19]	[kg PO3-4]	743.5	36.5		
C5	*Investment process*					
C5/I25	Stage of design/Knowledge of the design engineers	–	1	0.8	0.147	−0.178
C5/I26	Stage of design/Experience of the design engineers	–	1	0.1	0.147	
C5/I27	Stage of building/Number of offers – big contracts	number of offers	28	26	0.353	
C5/I28	Stage of building/Number of offers – local market small contracts	number of offers	5	2	0.353	

The "Building costs per m^2" is determined as the one most important criteria in the ranking of choice of road building technology in the selection process by AHP. Consistency ratios of the pairwise comparison matrixes are calculated less than 0.1. So the weights are shown to be consistent and they are used in the ranking process.

Finally, Hellwig method is applied to rank the choice of building road technology. The values of the weights with respect to criteria, calculated by AHP and shown in Table 2, can be used with Hellwig method. The data of the AHP with Hellwig method, can be seen on Table 7.

Table 7. The calculated values accordance with Hellwig method and with weights by AHP and the ranking.

No.	Pavement	Calculated values	Ranking
1	Bituminous	0.869	2
2	Concrete	0.883	1

Based on the analyses made with the use of the Hellwig method, it can be said that the obtained values of synthetic meters for asphalt and concrete pavements are quite close, which indicates that the technologies within the indicators taken for analyses in relation to the features technological- technical-usable ones are comparable.

Many factors have affected obtained results, among which we should mention statistic material, the set of criterions taken for testing and preferential analytics concerning these criterions.

Obtained results are convergent with results of analyses carried out by authors of this work with the use of Hellwig method of data normalization and weights taken subjectively based on the expert's knowledge [27].

4 Discussion

The using of AHP method in realization road works will make it possible to follow a multi-functional and sustainable development scheme combining the conditions of rational choice of building road technology. Choice of building road technology should be given comprehensive treatment as a process under which multidirectional activities are carried not only to improve the road structure of pavements but also to implement the concept of multifunctional development of roads. The above- mentioned activities include works related to improving the status of road management within the object of choice of pavements.

AHP is an effective decision making method especially when subjectivity exists and it is very suitable to solve problems where the decision criteria can be organized in a hierarchical way into sub-criteria. The results obtained are convergent with the findings of previous studies about using other methods by ranking of choice of road building, and which was presented in literature. This methodology can be applied to another area of research.

5 Conclusion

Due to strategic importance of choice of road building technology, extensive durability research of these both pavements are being done to cope with this problem.

When choosing a road building technology one cannot be limited to building costs alone, but it is necessary to also consider the costs of maintenance and exploitation some 30–40 years later. The main purpose when choosing the road building technology is to build the roads of such a quality that would enable their long time exploitation and usage. Based on the analyses made with the use of the Hellwig method, it is possible to state that both technologies, the asphalt one and the concrete one can be competitive since it leads to their progress and development. Considering the growth of traffic on the roads, concrete can be perceived as a technical and economic alternative to asphalt structures, which is confirmed by presented analyses.

Even the best designed and built concrete pavements will eventually wear out and require maintenance or rehabilitation. It is important to recognize that timely maintenance, or pavement preservation, can substantially extend pavement life and delay the need for rehabilitation.

Concrete pavements may also be used as overlays for either existing asphalt or concrete pavements. For each of the two existing pavement types, there are two overlay classifications based on whether the overlay is bonded to the existing pavement, or whether the bond is either ignored or prevented, and thus not considered in the design.

The integrated AHP and Hellwig method approach is proposed as an efficient and effective methodology to be used by decision makers on road sector in terms of its choice to deal with both qualitative and quantitative performance. The proposed methodology can also be applied to any other selection problem involving multiple and conflicting criteria.

The result of evaluation may help strategy makers of road/building sector, local municipalities, management of road agencies, etc. Also different other multi criteria techniques such as zero unification method, normalization method can be used for comparing the results.

References

1. Deja, J.: Polish roads, review of road and bridge technique (2003). (in Polish)
2. Szydło, A.: Road pavements made of cement concrete. Theory of dimensioning, implementation, Polski Cement (2004)
3. Szydło, A., Mackiewicz, P., Wardęga, R., Krawczyk, B.: Catalog of typical structures of rigid pavements, report by GDDKiA, Warszawa (2014)
4. Mackiewicz, P.: Thermal stress analysis of jointed plane in concrete pavements. App. Therm. Eng. **73**, 1167–1174 (2014)
5. Mackiewicz, P.: Analysis of stresses in concrete pavement under a dowel according to its diameter and load transfer efficiency. Can. J. Civ. Eng. **42**(11), 845–853 (2015)
6. Mackiewicz, P.: Finite-element analysis of stress concentration around dowel bars in jointed plain concrete pavement. J. Transp. Eng. **141**(6), 06015001 (2015)

7. Wu, Z., Mahdi, M., Rupnow, T.D.: Accelerated pavement testing of thin RCC over soil cement pavements. https://doi.org/10.1016/j.ijprt.2016.06.004
8. Harrington, D., Abdo, F., Adaska, W., Hazaree, C.: Guide for Roller-Compacted Concrete Pavements, National Concrete Pavement Technology Center, Institute for Transportation, Iowa State University (2010)
9. Macioszek, E., Lach, D.: Analysis of the results of general traffic measurements in the West Pomeranian Voivodeship from 2005 to 2015. Sci. J. Sil. Univ. Technol. Ser. Transp. **97**, 93–104 (2017)
10. Czarnecki, L., Woyciechowski, P., Adamczewski, G.: Risk of concrete carbonation with mineral industrial by-products. KSCE J. Civ. Eng. **22**(2), 755–764 (2018)
11. Czarnecki, L., Woyciechowski, P.: Concrete carbonation as a limited process and its relevance to CO2 sequestration. ACI Mater. J. **109**(3), 275–282 (2012)
12. Sztubecka, M., Bujarkiewicz, A., Sztubecki, J.: Optimization of measurement points choice in preparation of green areas acoustic map. Civ. Environ. Eng. Rep. **23**(4), 137–144 (2016)
13. Frost Durability of Roller Compacted Pavements, Canada, Portland Cement Association (2004)
14. Roller Compacted Concrete Pavements Design and Construction, US Army Corps of Engineers, Washington, D.C. (2000)
15. Guide Specification for Construction of Roller-Compacted Concrete Pavements, American Concrete Institute, Farmington Hills (2004)
16. Dobiszewska, M.: Waste materials used in making mortar and concrete. J. Mater. Educ. **39**(5–6), 133–156 (2017)
17. Beycioglu, A., Gultekin, A., Aruntas, H., Huseyin, Y., et al.: Mechanical properties of blended cements at elevated temperatures predicted using a fuzzy logic model. Comput. Concr. **20**(2), 247–255 (2017)
18. Macioszek, E., Lach, D.: Analysis of traffic conditions at the Brzezinska and Nowochrzanowowska intersection in Myslowice (Silesian Province, Poland). Sci. J. Sil. Univ. Technol. Ser. Transp. **98**, 81–88 (2018)
19. Macioszek, E., Lach, D.: Comparative analysis of the results of general traffic measurements for the Silesian Voivodeship and Poland. Sci. J. Sil. Univ. Technol. Ser. Transp. **100**, 105–113 (2018)
20. Macioszek, E.: Analysis of significance of differences between psychotechnical parameters for drivers at the entries to one-lane and turbo roundabouts in Poland. In: Sierpiński, G. (ed.) Intelligent Transport Systems and Travel Behaviour. Advances in Intelligent Systems and Computing, vol. 505, pp. 149–161. Springer, Cham (2017)
21. Skrzypczak, I., Radwański, W., Pytlowany, T.: Durability vs technical - the usage properties of road pavements. https://www.e3s-conferences.org/articles/e3sconf/abs/2018/20/e3sconf_infraeko2018_00082/e3sconf_infraeko2018_00082.html
22. Saaty, T.L.: Decision-Making for Leaders: The Analytic Hierarchy Process for Decision in a Complex World. Analytic Hierarchy Process Series, vol. 2, New Edition. RWS Publications, Pittsburgh (2001)
23. Saaty, T.L.: Creative Thinking, Problem Solving and Decision Making. RWS Publication, Pittsburgh (2005)
24. Saaty, T.L., Vargas, L.G.: The possibility of group choice: pairwise comparisons and merging functions. Soc. Choice Welf. **38**, 481–496 (2011)
25. Ali, N.H., Sabri, I.A.A., Noor, N.M.M., Ismail, F.: Rating and ranking criteria for selected islands using fuzzy analytic hierarchy process (FAHP). Int. J. Appl. Math. Inform. **1**(6), 57–65 (2012)
26. Aly, S., Vrana, I.: Evaluating the knowledge, relevance and experience of expert decision makers utilizing the Fuzzy-AHP. Agric. Econ. **54**(11), 529–535 (2008)

27. Skrzypczak, I., Radwański, W., Pytlowany, T.: Choice of road building technology - statistic analyses with the use of the Hellwig method https://www.e3s-conferences.org/articles/e3sconf/abs/2018/20/e3sconf_infraeko2018_00081/e3sconf_infraeko2018_00081.html
28. Len, P., Oleniacz, G., Skrzypczak, I., et al.: SGEM 2016, Book Series: International Multidisciplinary Scientific, GeoConference-SGEM, vol. II, pp. 617–624 (2016)
29. Asfal, J.: Difference Between Flexible And Rigid Pavement. http://www.engineeringintro.com/transportation/road-pavement/comparison-between-flexible-and-rigid-pavement-properties
30. Mrówczyńska, M., Sztubecki, J.: Prediction of vertical displacements in civil structures using artificial neural networks. In: 10th International Conference on Environmental Engineering, Vilnius Gediminas Technical University, pp. 1–7 (2017)

Multiple Criteria Evaluation of the Planned Bikesharing System in Jaworzno

Andrzej Bąk, Katarzyna Nosal Hoy[(✉)], and Katarzyna Solecka

Cracow University of Technology, Kraków, Poland
bak.andrzej@student.pk.edu.pl,
{knosal,ksolecka}@pk.edu.pl

Abstract. Bicycle traffic is increasingly being considered in transportation policies and bikesharing systems have become an inseparable element of transportation systems in many cities. This paper discusses bikesharing systems and their accessibility. The paper presents the results of an accessibility evaluation of the planned bikesharing system in Jaworzno (Poland). The variants of system development were evaluated using Multiple Criteria Decision Aid (MCDA) and QGIS software. With these methods, various evaluation criteria were covered related to system accessibility, including for people in various age groups, concerning accessibility of traffic generators and system integration with public transportation. The analyses resulted in a ranking of variants of system development, from the best to the worst in terms of the reviewed criteria.

Keywords: Bikesharing · Accessibility · Multiple Criteria Decision Aid · GIS

1 Introduction

Bike traffic continues to be incorporated into urban transportation policies; social awareness of the benefits derived from cycling and the need to develop bike infrastructure has been growing. In parallel, the idea of the sharing economy has been developing on a global scale [1], more and more systems for sharing vehicles, including bikesharing systems are being developed [2]. The first bikesharing system was set up in 1965 in Amsterdam [3]. However, this was different from today's systems. The system did not have rental outlets, there were no fixed fees. There were 50 white painted bikes made available free of charge for unrestricted use [2]. Similar (relying on citizens' sense of responsibility, without docking stations and control systems) first generation systems eventually failed due to theft and vandalism [3]. In second generation systems, bikes were rented from docking stations subject to prior deposit with no prior user registration required [4]. The third generation requires registration (in order to reduce costs resulting from vandalism), fees are charged directly to users' accounts and bikes may be rented and returned at any station. In the fourth generation, the system does not require stations and bikes are rented with the use of mobile devices, like smartphones or tablets or via an on-board computer fitted to the bike [2]. The number of bikesharing programmes operating globally has evolved over recent years, with over 800 cities in 2017 [5]. At the end of 2018, there were 59 city bike systems in Poland. Among them, only two

(Warsaw – a private system, and Kraków) are fourth generation systems. The others can be classified as third generation systems.

There are a number of factors affecting the usage level of bikesharing systems [3, 6]. Some of them relate to service quality, including costs and comfort of use, others refer to the existence of infrastructure for cycling and weather conditions [7]. However, the key factor supporting a high level of use is the location and distribution of bike stations [6, 8], which ensures high system accessibility.

Accessibility is a notion used in many areas of science and therefore it is understood and defined in many ways [9–11]. Gadziński defines accessibility of transportation systems as a "relation between social surrounding (local community characterised with a specific set of spatial behaviour) and the transport infrastructure and means of transportation" [12]. Faron presents an accessibility definition of the functional and spatial ease to reach destination subject to the location of the trip starting and end point (in terms of the trip length and type), with reference to various means of transportation and transportation infrastructure [13]. In terms of access on foot, she states that 400 m is the maximum acceptable distance that pedestrians are willing to cover to reach a destination (for instance a bikesharing station). Gehl states that the commonly acceptable walking distance is 500 m [14]. In terms of accessibility of public transport stops, walking distances should be a maximum of 300 m to 1000 m, depending on the type of surrounding developments while the acceptable walking distance is dramatically reduced when in excess of about 500 m [15].

With reference to bikesharing systems, the authors [16] claim that a good quality bikesharing system needs from 10 to 16 stations per square kilometre to ensure a distance of approx. 300 m between stations and thus, high accessibility to the service. Such systems, characterised by high station density, are typical for large cities and can be found for instance in Paris or Montreal [4]. On the other hand, high density of stations requires substantial investment, and according to [17] overcoverage may be detrimental to the success of the system because it increases maintenance costs. With respect to systems operated in smaller cities (<100,000 inhabitants), the results of a Swiss study confirm that in small cities network density also remains a critical variable to ensure higher levels of use [6]. When such a system is implemented, it is necessary to plan the number and locations of stations to ensure high accessibility in residential and other areas that generate much traffic (e.g. commercial, manufacturing) as well as place them in relation to public transport networks [18].

This paper discusses bikesharing and its accessibility. The paper presents the evaluation results of the planned bikesharing system in Jaworzno (Poland), subject to its accessibility. The variants of system development were evaluated using the Multiple Criteria Decision Aid (MCDA) and QGIS software. Section 2 presents the bikesharing concept for Jaworzno along with development variants, and Sect. 3 describes the methods applied to evaluate the variants. Section 4 shows the evaluation procedure and its results while Sect. 5 is a summary of the analyses.

2 The Concept of Developing a Bikesharing System for Jaworzno

The bikesharing system analysed in this paper is to be implemented in Jaworzno, a city in the south of Poland, in the Silesian province. It covers an area of 152.2 km^2 and is inhabited by 92,090 people [19]. Jaworzno has a Sustainable Urban Mobility Plan [20] which proposes the implementation of a bikesharing system to promote bike traffic. On 28 September 2018, the City of Jaworzno announced a tender for the delivery, installation and commissioning of a Jaworzno City Bike (JCB) system, combined with management, maintenance and comprehensive operation [21]. In its first stage, the system was to cover 20 stations, each with 10 bikes. The contractor was to provide 200 bikes (120 electric bikes and 80 traditional bikes). In the second stage, the system could be expanded by another 5 stations and 50 bikes. Additionally, during a workshop held with inhabitants in September, opinions were presented that the system should ultimately operate about 40 stations.

For the purposes of this study, six development variants of JCB have been developed, differing in the number and locations of stations. The number of stations was assumed on the basis of the tender documentation [21] and proposals by the inhabitants, and is 20, 25 and 40 respectively subject to system development stages. With respect to locations, two scenarios were reviewed: WA (covering primarily 23 station locations from the tender documentation and the location of the other 17 stations, set on the basis of the authors' concept) and WB (solely the authors' concept of station locations). In the authors' concept, station locations were planned in order to ensure the best accessibility from residential areas and the major commercial and industrial facilities. The locations were indicated on the basis of guidelines set forth in urban planning documents [20] and also after an inventory covering the spatial layout of buildings, locations of schools, kindergartens, major commercial facilities (large supermarkets, open-air markets, public offices), the existing and planned cycling infrastructure, the existing locations of bus stops and train stations. The developed JCB development variants are as follows:

- WA1 – covers 20 stations located as specified in the tender documentation;
- WA2 – covers 25 stations located as specified in the tender documentation and in the authors' concept;
- WA3 – covers 40 stations located as specified in the tender documentation and in the authors' concept;
- WB1 – covers 20 stations located solely on the basis of the authors' concept;
- WB2 – covers 25 stations located solely on the basis of the authors' concept;
- WB3 – covers 40 stations located solely on the basis of the authors' concept.

The JCB development variants presented above have been evaluated, taking into account the system accessibility.

3 Methods

To evaluate the accessibility of the planned bikesharing system (its variants), the Multiple Criteria Decision Aid (MCDA) methodology was applied. This is a methodology derived from operational research, alternatively called Multiple Criteria Analysis or Multiple Criteria Decision Aid Process [22, 23]. In the study of Zeleny [23] MCDA is defined as making decisions in the presence of many criteria/objectives, whereas in the work of Vincke [22], as solving complex decision problems where many, often opposing points of view must be considered. Multiple Criteria Decision Aid is a methodology that has been developing dynamically in recent years. It is used, among others, to solve transport related problems [24, 25].

According to Roy the basic attributes of the Multiple Criteria Decision Aid problems are [26]: a set A of solutions and a consistent family F of evaluation criteria. The set A of solutions is a set of decision objects or variants to be analysed and evaluated during the decision-making process. The set A of solutions may be defined: directly by listing all its elements (a sufficiently small set, a definite number of objects); and indirectly by defining properties that characterize all elements of set A or conditions limiting set A. The set A may be defined in advance and not subject to changes during the decision-making process or evolving (varying), i.e. subject to modifications during the decision-making process.

A consistent family F of criteria [26] is a set of criteria that meet the following requirements: exhaustiveness of the evaluation (contemplating all possible aspects of the problem under consideration); consistency of evaluation (based on proper determination of global decision preferences by the criterion); and the uniqueness of the criteria ranges. Each criterion present in the set F is a function f defined on the set A to evaluate the set A and representing the preferences of the decision maker (DM) in relation to a particular decision problem.

A multi-criteria decision problem is a situation in which, having a defined set A of solutions (variants) and a consistent family F of criteria, the DM seeks to [22]: (1) determine the subset of solutions (variants) considered to be the best for the family of criteria under consideration (selection problem); (2) divide the set of solutions (variants) into subsets according to certain standards (problem of classification or sorting); (3) rank the set of solutions (variants) from best to worst (problem of positioning or ranking).

For the evaluation of the variants of the bikesharing system in Jaworzno, one of the solutions for positioning methods was applied, namely the Compensation-Conjunction (CC) method. The method was proposed by Rudnicki [15]. This multiple criteria evaluation method expresses the principle of 'something for something'. Failure to meet one criterion can be compensated by a higher fulfilment of another criterion. The method includes the following elements of the evaluation procedure: (1) formulation of a list of criteria in a one-stage or multi-stage system; (2) determination of the weights of criteria; (3) determination of threshold criteria; (4) evaluation of the degree of fulfilment of individual criteria by the variants in question and the determination of the required minimum fulfilment for threshold criteria; (5) elimination of variants that do not meet the threshold criteria; (6) aggregation of partial evaluations, obtaining a global evaluation; (7) ordering variants by values due to the global rating indicator.

The global evaluation S_j of the j-variant is determined by the following formula [15]:

$$S_j = \sum_{i=1}^{n} w_i * s_{ij} \quad (1)$$

where:
s_{ij} – degree of fulfilment of the i-criterion in the j-variant (in percent, on a scale from 0% to 100%, where 0% means no compliance with the criterion, and 100% means complete compliance with the criterion or on a ten-point scale, where 1 means that the criterion is not met and 10 means complete fulfilment of the criterion),
  n – the number of criteria considered,
  w_i - weight of the i-criterion (non-rendered number, normalized), $w_i > 0$

$$\sum_{i=1}^{n} w_i = 1 \quad (2)$$

To determine the weights of the criteria and the degree of fulfilment of a given criterion, data from expert opinions are usually used. Based on the formula (1), the values of S_j are obtained on a scale of 1 to 10 points. The calculated values of S_j for particular considered variants allow their global (aggregated) quality to be evaluated. The higher the S_j value, the better the variant is considered to be.

In order to acquire the data related to the accessibility of each JCB variant as required for identifying the compliance of the approved criteria by each variant, a GIS tool was used – the state-of-the-art version of QGIS 3.4 Madeira [27], with long-term support assured. GIS tools are commonly applied in transport analyses providing for data collection, processing and visualisation, thus supporting decision processes [28–30]. With respect to using the data in studies on bikesharing systems, it is worth quoting research by Rybarczyk and Wu [31], who used GIS and a MCDA to determine optimal bicycle routes with multiple objectives, or research by García-Palomares et al. [4], who applied a GIS-based method to calculate the spatial distribution of the potential demand for bikesharing trips, locating stations using location–allocation models. In [32] GIS was applied to present distribution of modal shift to and from public transport, which results from bikesharing usage.

4 Multiple Criteria Evaluation of the Bikesharing System Variants

A multiple criteria decision problem was formulated as a multiple criteria problem for ranking variants of the bikesharing system in Jaworzno, providing for their accessibility. The evaluation covered 6 system development variants, as described in Sect. 2 (WA1 – WB3) and 9 evaluation criteria.

Considering the requirements defining a consistent set of criteria (exhaustiveness of evaluation, consistency of the evaluation and uniqueness of the criteria ranges of meaning [26]), the evaluation of the analysed variants relied on criteria related to the percentage of the population with specific accessibility of the system and the criteria related to accessibility to traffic generators (schools, commercial and industrial facilities) and providing for possibilities to integrate the system with public transportation. The criteria are presented in Table 1. In view of the acceptable walking distance suggested in [14] and [15] to the destination, the walking distance along existing roads for K1 to K8 criteria was assumed as 500 m. With respect to the criterion of transport integration K9, the walking distance was assumed as 100 m which is due to the need to ensure the most efficient transfer from a bus to a bike (or the other way round) which is the shortest distance between a bus stop and a bikesharing station.

Afterwards, weights were assigned to each criterion on the basis of an expert analysis. The experts (staff members of the Department of Transportation Systems at the Cracow University of Technology) assigned weighs to each criterion along a scale of 1 to 10, where 1 meant the lowest weight (unimportant criterion) and 10 – the highest weight (very important criterion). The weights assigned by the experts were averaged and standardised (Table 2).

Table 1. Evaluation criteria.

Criterion group	Criterion symbol	Criterion name
Criteria concerning population with service accessibility	K1	Percentage of population at pre-production age (0–14 years) within 500 m of a bikesharing station
	K2	Percentage of population at production age (15-64 years) within 500 m of a bikesharing station
	K3	Percentage of population at post-production age (65+ years) within 500 m of a bikesharing station
Criteria concerning accessibility to traffic generators	K4	Percentage of kindergartens within 500 m of a bikesharing station
	K5	Percentage of primary schools within 500 m of a bikesharing station
	K6	Percentage of senior high schools within 500 m of a bikesharing station
	K7	Percentage of services (discount stores, large supermarkets, open-air markets and public offices) within 500 m of a bikesharing station
	K8	Percentage of largest industrial facilities within 500 m of a bikesharing station
Integration criterion	K9	Percentage of bus stops within 100 m of a bikesharing station

The next step was to evaluate the degree of fulfilment of individual criteria by the variants in question. In this case, degrees of fulfilment of individual criteria are the values of the criteria identified as population percentage, traffic generators and bus stops (concerning the criteria specified in Table 1). The shares for each variant were identified with QGIS 3.4 Madeira and its extension QGIS Network Analysis Toolbox 3 (QNEAT3).

In the first stage, using demographic data from the National Census of Population in 2011, available in the form of a vectorial square grid with each side of 1 km, the number of inhabitants was estimated in all developments in the city. OpenStreetMap was the source of information on the developments. The polygons with the developments were presented as centroids (points). Using the embedded QGIS algorithms, the number of points was calculated in each square of the demographic layer. With the Point Sampling Tool (PST) plug [33] and a "field calculator", the number of inhabitants in each age group per each development was calculated by dividing the demographic data by the number of buildings therein.

Afterwards, a raster layer was generated with the cell dimension of 10 m per 10 m containing information on the distance of each raster cell to the closest bikesharing station. The distance was measured to previously coded city bike stations (points) along the road grid (lines) obtained from OpenStreetMap. The tolerance of measurement (the tolerance of grid shortcomings) was set at 5 m.

Again using the PST plug, the centroids representing buildings were assigned distances along the road grid to the closest bikesharing station. Similarly, distances were assigned to previously recorded points representing traffic generators: kindergartens, primary and senior high schools, commercial facilities, larger industrial facilities and bus stops. The details of distances were acquired on the basis of locations of points representing buildings, traffic generators and bus stops versus individual cells in the earlier generated raster layer. Afterwards, using the data process in QGIS, a spreadsheet was used to calculate the degree of fulfilment of criteria by individual JCB development variants.

The degree of fulfilment of individual criteria by the variants in question was presented as a percentage from 0% to 100%, where 0% meant no compliance with a criterion and 100% complete compliance with a criterion. Table 2 presents the results. An analysis of the results shows that the variants with the largest number of stations (WA3 and WB3), in particular the authors' concept of locations (WB3), were most closely compliant with the criteria relating to the percentage of population in pre-production and production age (K1, K2), percentage of kindergartens (K4), industrial facilities (K8) and bus stops with a specified accessibility to the system (K9). The closest compliance with the criteria related to the percentage of population in post-production age (K3) and senior high schools (K6) relates to the WB2 and WB3 variants, and the criterion of the percentage of primary schools (K5) – WA3, followed by WB2 and WB3. Criterion K7 (percentage of services with a specified accessibility to the system) is not complied with at all by the WA1 and WA2 variants and is best complied with by WB3.

Table 2. Weights of criteria and evaluation of degree of fulfilment.

Criteria	K1	K2	K3	K4	K5	K6	K7	K8	K9
Weights of criteria	0.07	0.15	0.07	0.06	0.10	0.12	0.15	0.14	0.14
Evaluation of degree of fulfilment [%]									
WA1	29	28	53	54	43	40	0	5	29
WA2	32	29	58	54	43	52	0	6	32
WA3	40	37	63	75	71	57	13	11	40
WB1	28	27	58	42	43	60	6	6	28
WB2	35	33	68	50	57	74	13	8	35
WB3	48	46	79	63	57	88	38	11	48

In the next stage, a threshold criterion was identified and the minimum acceptable degree of fulfilment of this criterion. This was detected as criterion K2 – the percentage of population in production age (15–64 years) within 500 m of a bikesharing station with the minimum value of degree of fulfilment being 25% (within the range of 0% to 100%). All the reviewed variants met the threshold criteria at least at the minimum acceptable level. At the final stage of the CC method, the global evaluation of the variants was calculated in line with the formula (1) presented in Sect. 3. The values of global variant evaluation and their final ranking from best to worst is presented in Table 3, whereas the applied procedure is presented in Fig. 1.

The results in Table 3 show that WB3 is the best variant in terms of the analysed evaluation criteria – the variant characterised by the largest number of stations (40 stations), the authors' concept of their locations and the largest percentage of total population within 500 m of a bikesharing station. The global evaluation for the variant was 51.36%. This means that the variant provides the best accessibility of bike stations in terms of the reviewed evaluation criteria. The second position is occupied by variant WA3, also characterised by the largest number of stations whose locations were determined partly on the basis of the tender documentation and partly on the authors' concept. The global evaluation for variant WA3 was 42.95%. The difference between WB3 and WA3 variants was 8.4% points. The third position is occupied by variant WB2 with 25 stations. The variant had a global evaluation of 39.64% or 3.3% points less than variant WA3. The last position in the ranking was occupied by variant WA1 with 20 stations, with the locations determined solely on the basis of the documentation. The global evaluation in that case was 28.66% which means that the variant offers low accessibility to bikesharing stations in terms of the reviewed criteria. Variants WA2 and WB1 can be treated as equivalent. Irrespective of the number of stations, higher positions are occupied by those variants whose locations were determined on the basis of the authors' concept.

Table 3. Global evaluation and final ranking of the variants.

Position in the ranking	Variant	Global variant evaluation [%]
1	WB3	51.36
2	WA3	42.95
3	WB2	39.64
4	WA2	31.49
5	WB1	31.34
6	WA1	28.66

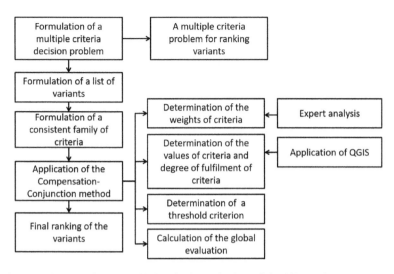

Fig. 1. Procedure applied in multiple criteria evaluation of the bikesharing system variants

5 Conclusions

Bikesharing systems have become an increasingly common element of city transportation systems. In 2019, the first system of the type is to be commissioned in Jaworzno. The plan was the inspiration for conducting the research described in this paper and the authors aimed at evaluating the development variants of the system, subject to various accessibility criteria. An additional objective was the use of free and open software (QGIS) as well as reliance on generally accessible information and materials. The multiple criteria analysis of six approved system development variants showed that the variants with the largest number of stations and locations set solely on the basis of the authors' concept of system development, subject to the approved evaluation criteria, are characterised with the best accessibility compared to the variants partly based on the locations specified in the tender and the author's future development vision. The results of the research may serve as a guideline for the authorities of the City of Jaworzno with respect to future development of the bikesharing system. The procedure applied in the research, due to the generally available methods and information, is universal and may be applied to similar analyses in other cities.

References

1. Ma, Y., Lan, J., Thornton, T., Mangalagiu, D.: Challenges of collaborative governance in the sharing economy: the case of free-floating bike sharing in Shanghai. J. Clean. Prod. **197**(P1), 356–365 (2018)
2. Shaheen, S., Guzman, S., Zhang, H.: Bikesharing in Europe, the Americas, and Asia: past, present, and future. Transp. Res. Record **2143**(1), 159–167 (2010)

3. Mateo-Babiano, I., Kumar, S., Mejia, A.: Bicycle sharing in Asia: a stakeholder perception and possible futures. Transp. Res. Proc. **25**, 4966–4978 (2017)
4. García-Palomares, J.C., Gutiérrez, J., Latorre, M.: Optimizing the location of stations in bike-sharing programs: a GIS approach. Appl. Geogr. **35**(1–2), 235–246 (2012)
5. Bao, L., Xu, C., Liu, P., Wang, W.: Exploring bikesharing travel patterns and trip purposes using smart card data and online point of interests. Netw. Spat. Econ. **17**(4), 1231–1253 (2017)
6. Audikana, A., Ravalet, E., Baranger, V., Kaufmann, V.: Implementing bikesharing systems in small cities: evidence from the Swiss experience. Transp. Policy **55**, 18–28 (2017)
7. Fishmana, E., Washington, S., Haworth, N., Watson, A.: Factors influencing bike share membership: an analysis of Melbourne and Brisbane. Transport. Res. A-Pol. **71**, 17–30 (2015)
8. Lin, J.R., Yang, T.H.: Strategic design of public bicycle sharing systems with service level constraints. Transport. Res. E-Log. **47**, 284–294 (2011)
9. Ingram, D.R.: The concept of accessibility: search for an operational form. Reg. Stud. **5**(2), 101–107 (1971)
10. Litman, T.: Measuring transportation: traffic, mobility and accessibility. ITE J. **73**(10), 28–32 (2003)
11. Puławska, S.: Accessibility contour measures in case study of Krakow. Logistyka **4**, 3207–3214 (2014)
12. Gadziński, J.: Functioning of a Local Transportation Systems versus Contemporary Urbanisation Processes. Example of the Poznań agglomeration. Bogucki Wydawnictwo Naukowe, Poznań (2013)
13. Faron, A.: Impact on selected factors of the city functional and spatial structure on assignment of transportation tasks. Doctoral dissertation. Cracow University of Technology, Krakow (2014)
14. Gehl, J.: Cities for People. Island Press, Washington (2010)
15. Rudnicki, A.: Quality of Urban Transportation. Polish Association of Transport Engineers and Technicians, Krakow (1999)
16. Gauthier, A., Hughes, C., Kost, C., Li, S., et al.: The Bike-share Planning Guide. ITDP Report (2013). https://www.itdp.org/wp-content/uploads/2014/07/ITDP_Bike_Share_Plann ing_Guide.pdf
17. Shu, J., Chou, M., Liu, Q., Teo, C.P., Wang, I.L.: Bicycle-sharing system: deployment, utilization and the value of re-distribution. Singapore: National University of Singapore-NUS Business School (2010). https://bschool.nus.edu.sg/staff/bizteocp/BS2010.pdf
18. Martens, K.: Promoting bike and ride: the Dutch experience. Transport. Res. A-Pol. **41**, 326–338 (2007)
19. Statistics Poland. http://stat.gov.pl/
20. Przedsiębiorstwo Projektowo-Usługowe INKOM s.c.: Update of the Sustainable Urban Mobility Plan of Jaworzno. Appendix to Resolution No. XVII/249/2016 of the City Council in Jaworzno of 30 March 2016 (2016)
21. Public Information Bulletin of the City Office in Jaworzno. http://bip.jaworzno.pl/Article/get/id,33579.html
22. Vincke, P.: Multicriteria Decision-Aid. John Wiley & Sons, Chichester (1992)
23. Zeleny, M.: Multiple Criteria Decision Making. McGraw Hill, New York (1982)
24. Galińska, B.: Multiple criteria evaluation of global transportation systems - analysis of case study. In: Sierpiński, G. (ed.) Advances in Intelligent Systems and Computing, vol. 631: Advanced Solutions of Transport Systems for Growing Mobility, pp. 155–171. Springer (2018)

25. Żak, J.: Design and evaluation of transportation systems. In: Sierpiński, G. (ed.) Advances in Intelligent Systems and Computing, vol. 631: Advanced Solutions of Transport Systems for Growing Mobility, pp. 3–29. Springer (2018)
26. Roy, B.: Multiple Criteria Decision Aid. Scientific and Technical Publishers, Warszawa (1990)
27. QGIS. https://www.qgis.org/en/site/
28. Lovett, A., Haynes, R., Sünnenberg, G., Gale, S.: Car travel time and accessibility by bus to general practitioner services: a study using patient registers and GIS. Soc. Sci. Med. **55**(1), 97–111 (2002)
29. McCray, T., Brais, N.: Exploring the role of transportation in fostering social exclusion: the use of GIS to support qualitative data. Netw. Spat. Econ. **7**(4), 397–412 (2007)
30. Kamruzzaman, M., Hine, J., Gunay, B., Blair, N.: Using GIS to visualise and evaluate student travel behavior. Appl. Geogr. **19**(1), 13–32 (2011)
31. Rybarczyk, G., Wu, C.: Bicycle facility planning using GIS and multi-criteria decision analysis. Appl. Geogr. **30**, 282–293 (2010)
32. Martin, E.W., Shaheen, S.A.: Evaluating public transit modal shift dynamics in response to bikesharing: a tale of two US cities. J. Transp. Geogr. **41**, 315–324 (2014)
33. Point Sampling Tool. https://github.com/borysiasty/pointsamplingtool

The Application of Bus Rapid Transit System in the City of Baquba and Its Impact on Reducing Daily Trips

Firas Alrawi[✉] and Yagoob Hadi

Urban and Regional Planning Center, University of Baghdad, Baghdad, Iraq
dr.firas@uobaghdad.edu.iq,
Yaaqoub.hadi.kl100b@iurp.uobaghdad.edu.iq

Abstract. This research deals with traffic jams in the city of Baquba and tries to find solutions that contribute to reducing the intensity of this congestion. The main objective of this research is to reduce traffic congestion in the city. Therefore, the study assumes that the application of the Bus Rapid Transit BRT will decrease the trips time in particular and the suffering associated with congestion in general. In this research, the BRT system proposed in the form of a two-way axis running from the eastern side of the intersections of Kanaan and Mafraq to the University of Diyala with a length of nearly 20 km.

The conclusions of this study present the most relevant outcomes: which is the possibility of converting people from using private transport to using public transport (BRT), which will reduce the use of private vehicles to about 50%, as well as it will shorten the time of trips and accessibility. Also, raise the level of road service to about 40%.

Keywords: Baquba · Transportation · Bus Rapid Transit BRT · Level of service

1 Introduction

The importance of this study is by the influence of its implications and functions around city and life quality. The transportation system considered as the lifeblood for towns, because of its precise role in linking different components of the city to each other and creating functional and spatial integration among them, as well as its visible impact on the social, economic and environmental aspects of the city. Most cities are growing as a result of population increase, as well as urbanization in developing countries, including Iraq. The development of transportation must accompany this growth as one of the main activities in human life since ancient times.

Changing to the (BRT express buses) system as a solution will help to reduce traffic problems, and it can change the current situation for the roads in the cities. It is one of the recommended options for medium-capacity transport systems [1]. These international systems have become highly efficient, low costs and many cities in the world are using the BRT, which is the backbone of urban development policy [2]. The BRT system provides many advantages in comparison to other transport systems; such as ease of implementation and management without high costs. The system provides high mobility speeds compared to other transport systems, high passenger capacity, safety, comfort levels, and reliability. The system can be among the sustainable transport systems with a low impact on the environment such as walking and Cycling, especially in case of using environmentally friendly transport modes [3].

1.1 Bus Rapid Transit (BRT)

A range of high-quality transport options based on bus use; many definitions can be found for the BRT. However, they all frame a bus-based system that simulates characteristics of high-performance urban capacity such as railways and metro at a lower cost [4].

The Institute of Transportation Policy and Development (ITDP), which located in New York, has emerged as one of the most prominent supporters of this technology. The BRT recognized as a "high-quality bus-based transport system" that provides fast, convenient, and cost-effective urban mobility by providing separate lanes [5].

So far, more than 150 cities have carried out some forms of the BRT system around the world, carrying an estimated 2 million passengers every day. These systems consist of 280 passageways around the world, with a total of 4,300 km, 6700 stations, and 30000 buses [6].

This system is currently being implemented with special features that are driving the development of the world following the success of its widespread use in Curitiba, Gauta, Mexico City, Istanbul, Ahmedabad, and Guangzhou. The high performance of this system of transport systems reflected in the development of these cities [7].

China has followed the steps of Latin America in building this system vigorously since it launched back in 2005 and over the past eight years has developed this system to add hundreds of kilometers to its network at a faster pace than any part of the world. Figure 1 shows the rapid development of BRT in China, and Fig. 2 shows the world's most widely used system [8].

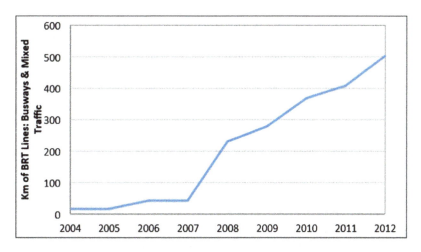

Fig. 1. Illustrates China's BRT development. Growth in China's BRT Network Lengths (in Kilometers, Two Directions): 2004 to 2012. (source: Institute 2013, p. 6)

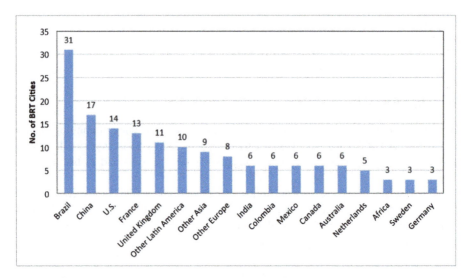

Fig. 2. Illustrates the most countries using BRT system. Number of Cities with BRT Systems, by National and Regional Settings. (source: Institute 2013, p. 5)

1.2 BRT and Population Density

In addition to factors such as income and vehicle ownership, urban density also has a substantial impact on passenger traffic, whether by bus or rail systems [9]. However, BRT in Latin America is much less intense than its Asian counterparts but tends to attract more customers. In general, it is evident in Latin America that other factors influence the use of this type of transport, where the ability to carrying the costs has played an essential and detailed role which affected positively to increase its effectiveness.

1.3 Global Experience the City of Curitiba

Although, Curitiba city is one of the highest cities in Brazil in terms of private vehicles ownership rates, the proportion of BRT system users was about 70% of the city population, which was positively reflected the low proportion environmental pollution. The BRT system in this city is fully integrated and works to ensure that all residents of the city, according to their requirements, reach the center of the city quickly and vice versa. The best evidence of the system's success is that while Curitiba's population has more than doubled since the system introduced in the 1970s, traffic has dropped by 30% [10, 11].

2 Case Study

Baquba is the center of Diyala province, about 50 km (31 miles) northeast of the capital Baghdad. It located on the Diyala River, with an estimated population of (342291) according to the enumeration of 2018, Fig. 3.

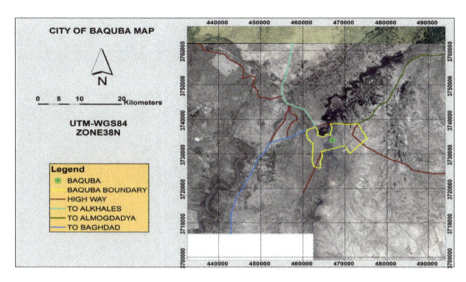

Fig. 3. Shows the location of the city center of Baquba for the province of Diyala. [12]

2.1 Traffic Situation in the City Center

The transport system in the city center of Baquba is characterized by a network of roads used for the movement of cars and pedestrian, which is an old network that has not been modernized compared to modern transport systems. Baquba also faced increasing in population, per-capita income, and car ownership without any change in the transport network. However, the worst thing is the absence of the public transport system that has negatively reflected in the performance of the transport network.

The streets in the city center drew a radial pattern towards the city's Suburban's, which represent a linear pattern of growth and the range of width between 20-40 m, it's linked with each other by sub-streets to reduce the impact of the long distance between the neighborhoods at the end of the belts and the city center, (CBD).

2.2 Data Collection

Through the field study which conducted, some points have elected as monitoring stations for data collection which distributed around the city center and its outer perimeter, which represented a contract and points of congestion as shown in the Fig. 4.

Data collection was carried out at these stations, as shown in Table 1, which shows that the small vehicles and medium buses passing through these routes in the city and causing heavy traffic congestion, as well as the environmental pollution associated with this congestion. The data represented one working day from 8 am to 2 pm.

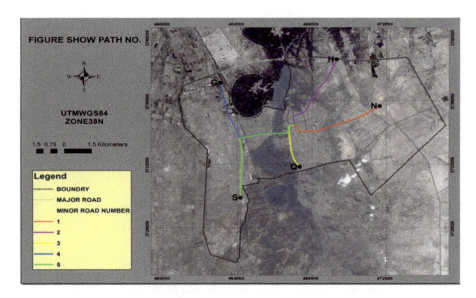

Fig. 4. Shows the locations of the monitoring points through which the data were collected. [12]

Table 1. Traffic volumes by types of vehicles passing by monitoring stations.

Tracks	Length/km	Monitoring points	Small vehicles	Medium buses
1	2.2	N	1196	253
2	2.7	H	1179	408
3	3.6	Q	3080	489
4	7.8	G	3004	323
5	2	S	2496	404

2.3 Data Analysis

The number of passengers was calculated through the number of vehicles collected in the field survey by multiplying its number by the car occupancy of each type Fig. 5. Moreover, Table 2 shows the types of vehicles and the capacity of each vehicle design. The number of passengers used for private cars is equal to the number of small cars occupancy, which is two passengers. If we have (1196 × 2), the number of passengers will be 2392 and so on the rest of the tracks. At the monitoring point (N), the number of passengers can be calculated by multiplying (253 × 9) passengers Medium-sized buses shall be the number of passengers for this type of vehicle is (2277), bringing the number of passengers.

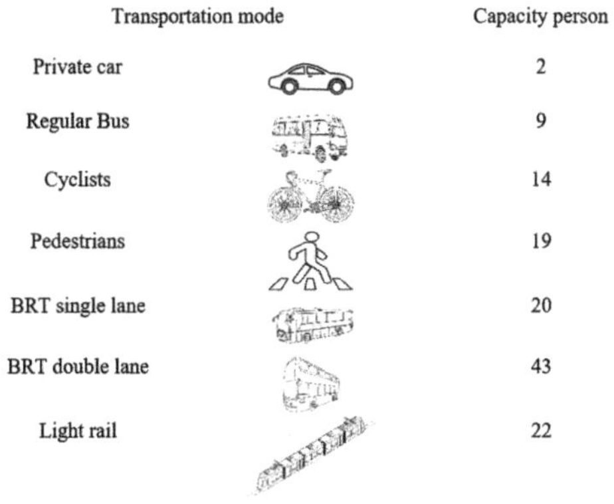

Transportation mode		Capacity person
Private car		2
Regular Bus		9
Cyclists		14
Pedestrians		19
BRT single lane		20
BRT double lane		43
Light rail		22

Fig. 5. Shows the absorptive capacity of transport systems. [14]

Through the (N) monitoring point is the sum of passengers by private vehicles and medium-sized buses, which amounts to (4669), and so on the rest of the tracks.

The environmental and mobility benefits of application BRT systems not coming from attracting riders who previously used a regular bus or commuted by train. However, from the new transit users, a study in the US estimated that 24 to 33% of riders served by new BRT systems are new transit users most having switched from the private car. [13] In Adelaide, 40 percent of those riding track guided buses were former motorists [4].

Table 2. Numbers of passengers for each type of vehicle according to their occupancy rate.

Tracks	Small vehicles	Small vehicles' passengers	Medium buses	Medium buses' passengers	Total of passengers
1	1196	2392	253	2277	4669
2	1179	2358	408	3672	6030
3	3080	6160	489	4401	10561
4	3004	6008	323	2907	8915
5	2496	4992	404	3636	8628

From the previous studies which discussed the percentage of passengers who were using private vehicles have switched to the use of BRT systems, which is 40% with effective public transport in those cities. So, we can say that around 50% of owners of private vehicles are expected to turn to this system at Baquba, which does not have effective public transport. From that, we can calculate the number of passengers based on the percentage expected above, which is equal to half the total number of passengers in privet vehicles, as shown in Table 3. The numbers of vehicles and passengers calculated as follows:

A - the number of small vehicles' passengers = the number of cars after reduction of 50% * 2
B - the number of medium buses' passengers = number of buses after reduction of 50% * 9
Total passengers in track = A + B

Table 3. Numbers of passengers served by the BRT system.

Tracks	Small vehicles	Small vehicles' passengers decreased by 50%	Medium buses	Medium buses' passengers decreased by 50%	Total of passengers
1	1196	1196	253	1139	2335
2	598	1179	408	1836	3015
3	598	3080	489	2201	5281
4	1591	3004	323	1454	4458
5	1296	2496	404	1818	4314

The calculation of the numbers of private vehicles and medium transport buses after reducing them to 50% as well as the addition of the number of cars that use these routes are as shown in Table 4.

Table 4. Numbers of vehicles using these tracks after applying BRT system.

Tracks	Small vehicles	Reduced cars no. by 50%	Medium buses	Reduced buses by 50%	Trucks
1	1196	598	253	127	576
2	1179	590	408	204	608
3	3080	1540	489	245	1396
4	3004	1502	323	162	1865
5	2496	1248	404	202	1290
Total	10955	5478	1877	940	5735

For the purpose of calculating the level of service for each track, the following equation should be applied for each path: Level of service = capacity/standard capacity
Road capacity = the number of vehicles on the road/number of survey hours (which is 6).

Table 5 shows the standard design for the roads and bridges in Iraq which used to estimate the level of service after determining the traffic volumes on each track, through comparing it with the standards levels of services illustrated in Table 6.

Table 5. Road design standards in Iraq.

Features	Freeways	Expressways	Arterials roads	Collectors roads	Local roads
Design power for km/h speed	100	100	80	60	40
The operation speed in km/h	80–100	60–80	40–60	30–50	20–30
Absorbed vehicle/hour	1800–2000	1000–1400	800–1200	600–900	500–700
Number of street traffic lines	4–8	4–8	2–6	2–4	1–2
The taboos of the streets	3–100 m	30–80 m	15–60 m	10–20 m	8–12 m

Table 6. The standards levels of services.

Level	Occupation ratio	Property
A	Less than 0.4	Where the flow is free, and the movement is high and low in traffic density
B	0.4–0.6	Action balanced on the street and high speeds
C	0.6–0.8	Stable Flow
D	0.8–0.9	Flow is approaching instability and relatively high speeds
E	0.9–1	Unstable flow, low speeds, and possible traffic jams
F	More than 1	Low flow with high delays with traffic paralysis

Table 7 shows the level of service in the selected tracks of study in the city of Baquba. It is clear from the table that all of these routes had a low level of service between C and F levels. This is due to the lack of public transport and the total dependence on private vehicles.

Table 7. Level of service in the local roads of Baquba before using the BRT system.

Tracks	Number of vehicles	No. of cars/hours of 6 monitors	Mayoralty of Baghdad capacity 500 cars/hour	Categories
1	2025	338	0.68	C
2	2195	366	0.73	C
3	4965	828	1.66	F
4	5192	865	1.73	F
5	4190	698	1.4	F

By following the same methodology, the improvement of the level of service can be measured on those tracks selected in the study if the BRT system is applied Table 8 shows the level of service in the local roads of Baquba city after the use of BRT system.

Through this simple comparison, it is possible to observe the magnitude of the effect of applying the BRT system on the level of service in the studied area. The service level has improved in all tracks although the level of service in some of these routes remains below the required level.

Table 8. Level of service in the local roads of Baquba after using the BRT system.

Tracks	Number of vehicles	No. of cars/hours of 6 monitors	Mayoralty of Baghdad capacity 500 cars/hour	Categories
1	1301	217	0.43	B
2	1402	234	0.47	B
3	3181	530	1.06	F
4	3529	588	1.18	F
5	2740	457	0.91	D

3 Conclusions

1. Through the study of the BRT testing of several cities, a wide range of key performance indicators have been found, that show the most successful solutions of the implementation of the system within cities such as the optimal use of traffic signals, traffic continuity, freight management services, enhanced network efficiency, reduce congestion, improve road safety, and enhance integration of transport systems.
2. Another benefit from the application of the BRT system on the results of traffic volumes, which indicates the presence of streets experiencing congestion during the day, especially at peak hours (to reduce the congestion level by applying the index of congestion reduction by 50%) on the traffic volumes of all streets of the study area for private vehicles and medium buses and reach results showing a reduction in the value of traffic volumes.
3. An improvement of the LOS for Baquba streets, after implementation of BRT system for some streets from level C to B and from level F to D. When comparing the total number of streets to the study area by service levels of 5 streets in both directions in the study area, Streets and roads in the study area. It was found that tracks 1 and 2, which are at the level of Service C, have been transformed into a better service level, level B, and the service level of track 5 from category F to level D has changed, while lanes 3 and 4 have remained at the same service level.
4. Through the follow-up to the global experiments, we find that the proportion of users of this system has exceeded the percentage expected in the project study phase. We believe that the levels of service in the city of Baquba will change again to better levels after the start of the system for the system possesses the specifications will be stimulating to use such as (Lower cost, faster access, newer technology, less pollution).
5. The study proves that there is a visible improvement in the service levels after the use of the BRT system, which will lead to reduced levels of pollution as a result of the reduction of the use of private vehicles.

References

1. Marte, C.L., Yoshioka, L.R., Medeiros, J.E.L., Sakurai, C.A., Fontana, C.F.: Intelligent transportation system for bus rapid transit corridors (ITS4BRT). Recent. Adv. Electr. Eng. **23**, 242–249 (2013)
2. American Association of State Highway and Transportation Officials (AASHTO): National Transportation Communications for ITS – The NTCIP Guide – NTCIP 9001 version v 04, Washington, p. 141, July 2009
3. Broaddus, A., Litman, T., Menon, G.: Transportation Demand Management, Eschborn, Germany, p. 4, April (2009)
4. Cervero, R.: The Transit Metropolis: A Global Inquiry. Island Press, Washington, D.C. (1998)
5. Wright, L., Hook, W.: Bus Rapid Transit Planning Guide, 3rd edn. Institute for Transportation & Development Policy, New York (2007)
6. Cervero, R.: Bus Rapid Transit (BRT): an efficient and competitive mode of public transport. IURD Working Paper 2013–01 (2013)
7. Suzuki, H., Cervero, R., Iuchi, K.: Transforming Cities with Transit: Transit and Land Use Integration for Sustainable Urban Development. World Bank, Washington, D.C. (2013)
8. Pushkarev, B., Zupan, J.: Public Transportation and Land Use Policy. Indiana University Press, Bloomington (1977)
9. Ewing, R., Cervero, R.: Travel and the built environment: a meta-analysis. J. Am. Plan. Assoc. **76**(3), 265–294 (2010)
10. Curitiba: Story of a City. https://www.yesmagazine.org/issues/cities-of-exuberance/curitiba-story-of-a-city
11. How Curitiba's BRT stations sparked a transport revolution – a history of cities in 50 buildings, day 43. https://www.theguardian.com/cities/2015/may/26/curitiba-brazil-brt-transport-revolution-history-cities-50-buildings
12. Municipality of Baquba, Aerial photograph of the city of Baqubah, MSD extension (2018)
13. Prayogi, L.: Technical characteristics of bus rapid transit (BRT) systems that influence urban development, dissertation of Master of Urban Planning, The University of Auckland, 30 October 2015
14. Litman, T.: Transportation Cost and Benefit Analysis II – Congestion Costs, Victoria Transport Policy Institute VTPI, United States of America, p. 6 (2012)

The Application of Lithium-Ion Batteries for Power Supply of Railway Passenger Cars and Key Approaches for System Development

Viacheslav Bondarenko[1], Dmytro Skurikhin[1], and Jerzy Wojciechowski[2(✉)]

[1] Railway Cars Department, Ukrainian State University of Railway Transport, Kharkiv, Ukraine
bonvya@ukr.net, skurikhin@i.ua
[2] Faculty of Transport and Electrical, Kazimierz Pulaski University of Technology and Humanities in Radom, Radom, Poland
j.wojciechowski@uthrad.pl

Abstract. At present, the main source of electric power of the modern autonomous power supply system of the railway car is a generator driven by the gear unit from the axle of the wheel-set. Rechargeable batteries are used as a backup and emergency power source in all types of passenger cars. At the same time, car power supply systems with tradition generators and batteries have significant disadvantages. Existing car batteries have capacity limitations due to their weight and dimensions. Storage batteries take a significant part of the underbody space. The purpose of this article is investigating the possibility of using lithium-ion batteries for the power supply of passenger cars. The application of lithium-ion batteries for the power supply of passenger cars was considered using the example of the lithium-ion modules of the Nissan Leaf electric vehicle, which now is the most popular in Ukraine and other countries of the world.

Keywords: Railway · Passenger car · Power supply systems · Generator · Lithium-ion batteries · Electric power consumption · Capacity · Specific energy

1 Introduction

The main source of electric power of the modern autonomous power supply system of the railway car is a generator driven by the gear unit from the axle of the wheel-set. Rechargeable batteries are used as a backup and emergency power source in all types of passenger cars. The purpose of this article is investigating the possibility of using lithium-ion batteries for the power supply of passenger cars [1–5].

The following tasks should be identified to achieve the purpose:

- proposing approaches to the use of storage batteries of high specific energy on cars,
- calculating the electric power consumption of the car during the run,
- substantiating the choice of perspective batteries on cars,
- determining the number of modules, capacity and weight of the battery pack for the passenger car with the highest energy consumption, and consequently proposing the method of providing batteries for cars of different types,

- considering the possibility of transferring cars to autonomous (mixed) power supply systems without generator (battery-powered supply systems).

In our opinion, car power supply systems with generators remain insufficiently effective in operation and, despite existing improvements, do not fully correspond to the state-of-the-art level of development of science and technology.

There are two approaches can be identified in using lithium-ion storage batteries on passenger cars:

- Storage batteries of high specific energy are used as the main power source in car power supply systems (prospective) without generators. With this approach, regular batteries and generators with drives are replaced with batteries with increased specific energy and capacity.
- Storage batteries of high specific energy are used as a backup power source in all typical (traditional) power supply systems for cars. With this approach, when regular car batteries are replaced with prospective batteries, the standard capacity is preserved and the weight is reduced.

2 Power Supply for Vehicle Equipment

The electric power supply to the car equipment is a complex process influenced by multiple factors and limitations, such as: the power of car consumers, the size and weight of electric power sources, the cost of generated electric energy, the autonomy of cars, the electrification of railways, etc. [1–6]. By today, the development of passenger car design and electric equipment has resulted in the following main power supply systems: autonomous (mixed) and centralized (Fig. 1).

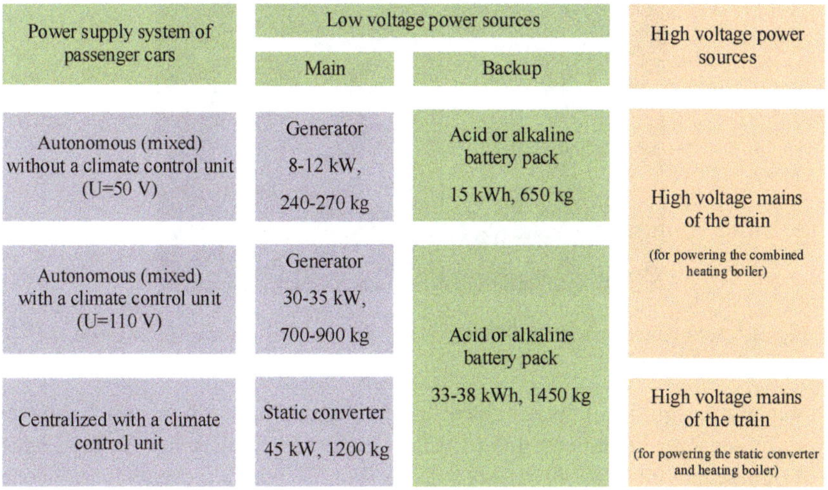

Power supply system of passenger cars	Low voltage power sources		High voltage power sources
	Main	Backup	
Autonomous (mixed) without a climate control unit (U=50 V)	Generator 8-12 kW, 240-270 kg	Acid or alkaline battery pack 15 kWh, 650 kg	High voltage mains of the train (for powering the combined heating boiler)
Autonomous (mixed) with a climate control unit (U=110 V)	Generator 30-35 kW, 700-900 kg	Acid or alkaline battery pack 33-38 kWh, 1450 kg	
Centralized with a climate control unit	Static converter 45 kW, 1200 kg		High voltage mains of the train (for powering the static converter and heating boiler)

Fig. 1. Typical power supply systems of modern passenger cars (source: own elaboration)

The main source of electric power of the modern autonomous (mixed) power supply system of the car is a generator driven by the gear unit from the axle of the wheelset.

Car power supply systems with generators have significant disadvantages, such as:

- high cost of electricity produced by the generator,
- creation of extra resistance to the train movement, especially when the power of car consumers increases,
- the need for separate departments for maintenance and repair of generators and their drives,
- low operational reliability of drives of generators, etc.

The existing ways to increase the efficiency of power supply systems with generators include the use of two generators with drivers from the end of the crank axis, mounted on both bogies of the car [7]. Generators work independently, one of them is designed exclusively for powering the climate control system. This design has its advantages vs. the existing autonomous systems, however, it still has the deficiencies specific for generating systems, and thus this design has not been used in practice.

In our opinion, car power supply systems with generators remain insufficiently effective in operation and, despite existing improvements, do not fully correspond to the state-of-the-art level of development of science and technology.

At present, rechargeable batteries are used as a backup and emergency power source in all types of passenger cars. KM260R alkaline batteries consisting of 40KM300P and 90KM300R batteries as well as L02300G and L02370G sections with acid gel electrolyte in 28L04300G and 28L04370G batteries are used in modern cars (Fig. 2). The specifications of these batteries are available on the manufacturer's official website [8].

Fig. 2. Acid battery pack 28L04370G under the body of passenger car (source: [8])

Despite the significant improvement of the characteristics of existing car batteries, they have capacity limitations due to their weight and dimensions. Storage batteries take a significant part of the under-body space (Fig. 3), which prevents from increasing the capacity and using them as the main sources of electric power for car consumers.

In this connection, the important tasks in the use of passenger car batteries include the search for ways to reduce their weight and dimensions preserving the constant capacity, ensuring insensitivity to different modes of operation, minimization of maintenance operations for batteries, etc.

Fig. 3. Under-body space of passenger car taken by battery pack (source: [9])

3 Main Part of the Study

Two approaches can be identified in using lithium-ion storage batteries on passenger cars:

- Storage batteries of high specific energy are used as the main power source in car power supply systems (prospective) without generators. With this approach, regular batteries and generators with drives are replaced with batteries with increased specific energy and capacity. The required autonomy of the car is determined on the basis of the consumption of electric power by car consumers between railway stations where battery chargers are available.
- Storage batteries of high specific energy are used as a backup power source in all typical (traditional) power supply systems for cars. With this approach, when regular car batteries are replaced with prospective batteries, the standard specific energy and capacity is preserved and the weight is reduced.

Methodology [6] was used to determine the amount of electricity consumed by the equipment of passenger cars during the train run. The most energy-consuming 61–779 car in the Kharkiv-Kyiv train with a duration of a run of approximately 8 h was chosen for calculation. The output data and calculated values are given in Table 1.

The calculation results show that the power consumption in the summer is almost three times higher than in winter due to the operation of the climate control system in the cooling mode. In winter, the climate control system is powered by a combined electric and coal boiler of the car, which, unlike the air conditioner, is powered by the high voltage mains of the train or solid fuel burnt in the boiler furnace. Therefore, the electricity consumption was calculated for the summer operating.

Table 1. Results of calculation of electricity consumed by the electrical equipment of the car during the run (source: own elaboration).

Consumer	P_H [kW]		K_{use}^a	Duration of operation, h	Consumed electricity [kWh]
	Summer	Winter			
General ventilation fan motor, P_v	2	2	0,84–0,9	6,72	13,44
Condenser fan motor, P_{cv}	1,4	–	0,73–0,8	6	8,4
Compressor motor, $2P_c$	13,6	–	0,6–0,75	4,8	65,28
Heating pump motor, P_{hp}	–	0,1	0,3	–	–
Electric heater of the air conditioner, P_{ehc}	6	–	0,85–0,9	–	–
Lighting system, P_{ls}	0,8	0,8	0,8–0,85	6,4	3,2
Laptop, charger, electric razor converters, P_{con1}	1	1	0,75–0,78	6	6
TV, video set converters etc., P_{con2}	0,25	0,25	0,75–0,78	6	1,5
Refrigerator converter, P_{con3}	0,2	0,2	0,9–0,95	7,2	1,44
Microwave oven and vacuum cleaner converter, P_{con4}	1,8	1,8	0,3	2,4	4,32
Electric water heater, P_{wh}	2,5	2,5	0,27	2,16	5,4
Boilers, P_{boil}	1,3	–	0,6	4,8	6,24
Electric heaters of waste and feeding pipes, P_p	–	0,7	0,06–0,18	–	–
Vacuum WC, P_{WC}	0,6	0,6	0,4	3,2	1,92
Switching cabinet, P_{SC}	0,55	0,55	0,9–0,95	7,2	3,96
Total[b]	26	10,5	–	–	121

[a] Duration of operation of the equipment during the run is calculated using the coefficient of use k_{use}.
[b] Since the air cooling ($P_{cv} + 2P_c$) and heating (P_{ehc}) systems of the air conditioner never run simultaneously, the greater of two powers is taken into account to determine the total power.

The total electricity consumption for an 8-hour car run was 121 kWh, but this estimated value needs to be verified due to the following reasons:

- rated power of devices was chosen for the calculation, although they not always operate in the stable nominal mode during the run,
- fluctuations in performance of climate control systems depending on the time of day and other external conditions were not taken into account during the calculation.

It should also be noted that only 38% of the passenger cars of Joint Stock Company «Ukrainian Railway» (JSC «Ukrzaliznytsia») consume about 121 kWh, the rest of the cars are not equipped with any climate control systems and have several times less power of car consumers.

Electricity metering in cars is performed with Train automated information and diagnostic system (PAIDS "VID") (Fig. 4). PAIDS "VID" together with new Acoustic on-board monitoring system for passenger cars was reviewed here [10]. The real-time collection and processing of statistics on electric power consumption by the equipment of passenger cars run requires additional research.

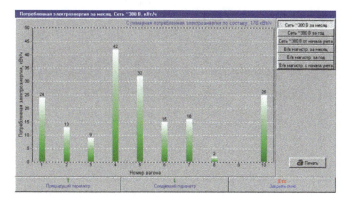

Fig. 4. Data of metering the electricity consumed for each car - a window of PAIDS software application (axis x - car number, axis y - energy consumption; in Russian) (source: [10])

The following factors should be considered in substantiating the choice of a promising battery for the power supply in passenger cars:

- specific energy, weight and dimensions,
- insensitivity to different operating modes and conditions,
- service life,
- modular design,
- market offer and price.

In order to study the issue as a whole, the global transport trends should be reviewed, where lithium-ion (li-ion) batteries have been significantly developed and widely used recently. The most powerful of them are used as the main sources of power in electric buses, electric trucks, and even ferries. However, the most commonly lithium-ion rechargeable batteries are used as the main source of power in electric and hybrid cars (Fig. 5).

The Japanese electric car Nissan Leaf has been acknowledged as one of the most popular cars in the world. The total sales of Nissan Leaf recorded in January 2019 were 363,940 cars. The largest sales markets were the US, Japan and Europe.

Fig. 5. Battery-powered vehicles (source: [11])

According to the statistics published by the Ministry of Infrastructure, as of late 2018, 11 thousand cars were registered in Ukraine. According to the diagram (Fig. 6), Ukrainians bought 2532 electric cars in 2017 and 7177 in January – November (I-XII) 2018.

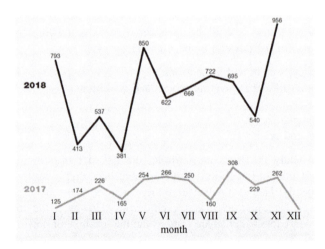

Fig. 6. Growth of the electric car market in Ukraine (source: [9])

Besides, since 2018, electric vehicles can be purchased free of VAT, excise and import duties. The Verkhovna Rada adopted in general bill No. 6776-d on amendments to the Tax Code, thus abolishing the value-added tax and the import excise for electric cars in Ukraine.

Nissan Leaf cars are the most popular in Ukraine and in the world, which allows forecasting a decline in the market value of storage batteries vs. other electric vehicles, both in the primary and after-markets. After analyzing the after-market based on the data of electronic publications and market platforms, the following change of prices for batteries of Nissan Leaf electric cars could be identified (Table 2).

Table 2. Change of prices for Nissan Leaf batteries in the after-market (source: [11]).

Year the battery was sold	Condition	Cost of the battery pack 24 kWh, USD	Specific cost of 1 kWh, USD
2017	Used (84% capacity)	4–5 thousand	167–208
2018	Used (86% capacity)	2.5–3.1 thousand	104–155
2019	Used (83% capacity)	2.3–2.6 thousand	96–108

The data above suggest that prices for lithium-ion batteries of electric cars decrease with time and already in the early 2019 1 kWh costs about $100 in the after-market.

Nissan Leaf lithium-ion batteries have a generally good design, which is proved by testing and operation. The following advantages should be noted in order to determine the possibility of using these batteries in railway passenger cars:

- low weight and dimensions of the elements due to their high specific energy – 130 Wh/kg vs. 25 Wh/kg of regular car batteries,
- modular design allowing to achieve the necessary voltage and capacity,
- similar operational conditions of batteries of railway cars and electric vehicles,
- consumers of the car and electric vehicle are similar in nature – electric drives, climate control systems, electronics and automation equipment,
- after 5 years of operation, the capacity of the battery decreases by an average of 15%, and then the process of degradation slows down.

The battery pack of the Nissan Leaf electric car consists of 48 separate modules, the specification of which is shown in Fig. 7. A package of such modules of the corresponding total capacity is proposed for use on cars as the main source of electric power. This approach allows applying the technologies that have already been tested and improved by Nissan leading to significant reduction of the cost of upgrading the power supply system of cars.

A similar example of using lithium-ion batteries is proposed for the electric bus Yoka ECO Bus, which has three battery packs, three electric motors and an inverter from the Nissan Leaf electric car [11].

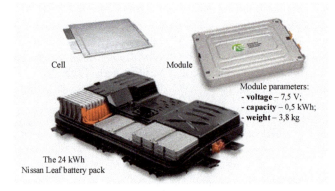

Fig. 7. Modular design of a lithium-ion battery pack of Nissan Leaf [11]

The capacity of a lithium-ion battery pack for a prospective car power supply system without generator is rated to be sufficient for one run of the train. Charging is proposed to be performed at the points of train formation and overnight stay from the external mains. Provided the railway line is electrified and a charger is available in the each car, lithium-ion batteries can be charged centrally from the high voltage mains of the train.

The estimated calculation of the number of lithium-ion battery modules of the car to form a capacity of 120 kWh and the connection diagram of separate modules in the section are shown in Fig. 8.

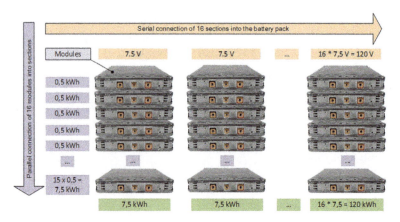

Fig. 8. Determination of the number of modules in the lithium-ion battery pack of the passenger car (source: own elaboration)

Using the above methodology, the necessary capacity for passenger cars of different types and power supply systems can be formed from the modules of a lithium-ion battery, and after transferring to the power systems without generator, the diagram of Fig. 1 can be represented as in Fig. 9.

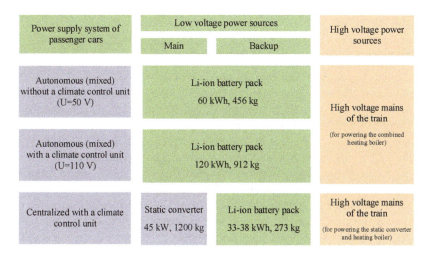

Fig. 9. Perspective power supply systems for passenger cars based on lithium-ion storage batteries (source: own elaboration)

The proposed power supply system of the passenger car should have a freely configurable battery pack with the possibility of quick change of the number of modules to form the required capacity at the standard output voltage.

A "battery storage car" can be proposed as an alternative to the outdated "generator car" for the centralized power supply of the train, which instead of diesel generating units contains lithium-ion rechargeable systems to store electricity.

It should be noted that cars with battery-powered supply systems have certain capacity limitations, there for the following measures should be focused on during re-equipment:

- developing and equipping the cars with the electricity metering devices,
- upgrading the consumer control system, including the provision of an energy saving mode ("eco-mode"),
- using energy-efficient climate control systems, including air conditioners with the heat pump mode,
- using LED lamps for lighting systems, etc.

The concept of the car battery-powered supply systems is consistent with the renewable energy technologies that have been developing intensively recently. One of such technologies for the railway transport is using solar panels on the roof of cars as an extra source of electricity. The Railway Cars Department of the Ukrainian State University of Railway Transport has developed a patent for "The method of alternative power supply of car of railway rolling stock from the photovoltaic system", which is relevant for use in the car facilities from an economic and environmental point of view [12].

4 Conclusions

The following conclusions can be summarized:

(1) Lithium-ion storage batteries should be considered as prospective sources of power supply of passenger cars. High specific energy, low self-discharge, absence of necessity of maintenance and other advantages have led to their common use in many types of modern electric vehicles.
(2) The approaches to the use of lithium-ion batteries in power supply systems of passenger cars were proposed and the advantages of such systems were determined. In prospective car battery-powered supply systems, where rechargeable lithium-ion batteries of high capacity are the main source of power, the need to use generators and their drives is eliminated. In the conventional power supply systems for cars, where lithium-ion storage batteries will have a nominal capacity and will be a backup power supply, the weight of the electric equipment of the car will be reduced from 500 to 1500 kg.
(3) The total power consumption of the car with the biggest number of electric users during the run was estimated to be further used as the basis for determination of the number of modules, capacity and weight of lithium-ion batteries proposed as the main source of energy supply for cars. The use of lithium-ion batteries for the power supply of cars was considered using the example of the lithium-ion modules of the Nissan Leaf electric car, which now is the most popular in Ukraine and other countries of the world.

References

1. Calderaro, V., Galdi, V., Graber, G., Capasso, A., Lamedica, R., Ruvio, A.: Emanagement of auxiliary battery substation supporting high-speed train on 3 kV DC systems. In: International Conference on Renewable Energy Research and Applications (ICRERA), Palermo, pp. 1224–1229. IEEE (2015)
2. Gu, Q., Tang, T., Cao, F., Song, Y.: Energy-efficient train operation in urban rail transit using real-time traffic information. IEEE Trans. Intell. Transp. Syst. **15**(3), 1216–1233 (2014)
3. Teshima, M., Takahashi, H.: Lithium ion battery application in traction power supply system. In: 2014 International Power Electronics Conference (IPEC-Hiroshima 2014 - ECCE-ASIA), Hiroshima, pp. 1068–1072 (2014)
4. Steiner, M., Klohr, M., Pagiela, S.: Energy storage system with ultracaps on board of railway vehicles. In: European Conference on Power Electronics and Applications, Aalborg, pp. 1–10 (2007)
5. Aiguo, X., Shaojun, X., Yuan, Y., Xiaobao, L., Huafeng, X., Jingjing, F.: An ultra-capacitor based regenerating energy storage system for urban rail transit. In: Energy Conversion Congress and Exposition, pp. 1626–1631. IEEE (2009)
6. Bondarenko, V.V., Obuhovs'kyj, V.V., Shataev, V.M.: Elektrychne obladnannja vagoniv [Electric equipment of railway cars]. UkrSURT, Kharkiv (2015). (in Ukrainian)

7. Kuksa, J.J.: System of Power Supply of Passenger Railroad with Two Undercar Generators. Ukrainian patent, No. 17040 (2006). http://base.uipv.org/searchINV/search.php?action=viewdetails&IdClaim=97645
8. Sayt zavodu "VLADAR Har'kovskij akkumuljatornyj zavod" [Site of Kharkiv battery factory] (in Russian). http://www.vladar.com.ua/ru/products.php
9. https://mtu.gov.ua
10. Makarenko, V.N., Bondar, S.I., Gambarjan, G.R., Bandura, I.N.: Sistema distancionnogo monitoringa zheleznodorozhnyh poezdov "Udalennyj "Vid" [The system of remote monitoring of railway trains "Remote" View"]. Zaliznychnyy transport Ukrayiny – Railway Transport of Ukraine **3**, 64–66 (2005). (in Russian)
11. Nissan LEAF technology to be used in electric bus test in Japan (in English). https://newsroom.nissan-global.com/
12. Bondarenko, V.V.: Method for Alternative Power Supply of Carriages of Rail Rolling Stock from Photoelectric System. Ukrainian patent, No. 101017 (2013). http://base.uipv.org/searchINV/search.php?action=viewdetails&IdClaim=183378

Time Parameters Optimization of the Export Grain Traffic in the Port Railway Transport Technology System

Oleg Chislov[1], Taras Bogachev[2], Alexandra Kravets[1],
Victor Bogachev[1], Vyacheslav Zadorozhniy[1(✉)], and Irina Egorova[1]

[1] Rostov State Transport University, Rostov-on-Don, Russia
{o_chislov,kravec_as,zadorozniy91}@mail.ru,
bogachev-va@yandex.ru, zezezel986@list.ru
[2] Rostov State University of Economics, Rostov-on-Don, Russia
bogachev73@yandex.ru

Abstract. The author's approach to the issues of time parameters optimization of the distribution of freight traffic in the regional railway transport and technological system, taking into account infrastructure indicators and economic and geographical parameters of cargo accumulation, is presented. A transportation process model has been formed, which is a two-criterion problem of integer mathematical programming. The procedure for finding the optimal plan of grain cargo flows by two time criteria was brought to numerical results using software.

Keywords: Multimodal transportation · ABC analysis · Grain storage railway station · Port station · Distribution of export grain traffic · Transport technology system · Optimization criterion · Traffic simulation

1 Introduction

The issues of regulation of multimodal transportation of goods to port stations are investigated in many papers [1–8]. For example, in [4] the main factors that need to be considered when choosing the locations of internal ports for their effective functioning are highlighted. In [5] operations performed by internal ports are studied, and the life cycle of these operations is determined. Also, their role is explored in international transport corridors and suggests ways to function effectively on Canada-connected transcontinental rail routes, especially in Asian directions. The analytical model of routing of cargo traffic, based on the principles of the theory of stocks and explaining how closely the organization of freight traffic is related to the characteristics of the relationship of the sender and the recipient is presented in [7]. We note that in the distribution of cargo flows there are collisions of commercial interests, not only competing modes of transport, but also entities of the transportation process within each of these types and even individual transport enterprises.

2 Actuality of the Time Minimization Problem for the Grain Export Traffic

The results of the analysis of indicators of transportation processes carried out in various countries indicate incomplete satisfaction of the transport services needs arising in the freight transportation market, as well as the loss of time resulting from the irrational distribution of freight traffic. Thus, there is no doubt about the relevance of the development of new mathematically grounded approaches to the issues of optimizing the distribution of cargo flows, in particular, railway, both existing and predictable.

In this paper, we construct a general mathematical model of the transportation process in considered region, which is a two-criterion problem of integer programming. Time parameters are considered as criteria. Thus, first of all, this model is applicable to transportations goods, for which the time factor is the main one. As far as we know, the method proposed below has not been considered in the researches devoted on freight traffic.

Further we will deal with such valuable (and traditional for the export of Russia) cargo as cereals. We pay special attention to the seasonal fluctuations in world prices for these raw materials, which are the most important regulator of demand and the incentive for its transportation.

During periods of substantial increase in grain prices on the external market, the number of applications for its export supplies from the southern regions of Russia increases sharply. In view of the above, it seems appropriate that all participants in the transportation process should be prepared for the indicated price fluctuations with the help the theoretically sound plan for the optimal distribution of freight traffic under conditions increased demand for the corresponding rail traffic.

3 Selection of Cluster-Forming Grain Storage Railway Stations Based on ABC Analysis

It is necessary to assess the infrastructure of transport system capabilities in order to organize an efficient grain cargoes supply [9]. Features of the infrastructure determine the temporary storage of goods in warehouses, the limit of grain loading and unloading intensity, the transportation period and other logistic chains parameters. It is necessary to identify the grain storage (storage capacity), on the basis of which it is possible to create transport grain clusters for the organization of grain cargoes export transportation to solve the above task. Grain storage railway stations have a number of technical characteristics, of which the most important for the organization of export cargo traffic, are: storage time, capacity storage and productivity of loading and unloading devices.

For the traffic routes with export grain, taking into account the interests of the cargo owner, the owner of the rolling stock, the carrier, the high productivity of the loading grain storage railway stations is necessary. We will select cluster-forming grain storage railway stations using modified ABC analysis [10].

Within the Black Sea region there is a large number of capacities grain storage of various capacities at railway stations. The variety, quantity and dispersion of grain storage railway stations its difficult when choosing the most effective points of cargo accumulation for subsequent transportation to the ports of the Black Sea. From the point of view influence on the traffic flow, their functions are not defined.

The estimated sample for the ABC analysis (Table 1) consists of the characteristics of 102 grain storage railway stations located in the Black Sea region, the so-called Big Black Sea Coast, having railway access roads, with a capacity from 42 to 300 thousand tons, with a capacity of 360–3180 tons/day. High productivity grain storage railway stations have a significant effect on the intensity and density of the transport network traffic flow with their low concentration.

ABC analysis was carried out for identify the most promising from the point of grain storage railway stations transport logistics view, for their ranking (Fig. 1).

Table 1. ABC-analysis of grain storage railway stations at the North-Caucasus Railways by productivity (fragment).

Object (station)	Classification criterion, Q, tons per day	Specific gravity, qj, persent	Growing total Sqj, persent	Classification group
Object 1 (Beloglinskaya)	3180	3.43	3.43	A
Object 2 (Zelenokumsk)	2160	2.33	5.76	A
Object 54 (Sulin)	720	0.77	73.57	B
Object 55 (Andreedmitrievka)	600	0.65	74.22	C
...	...	...	...	...
Object 102 (Khotunok)	180	0.19	100	C

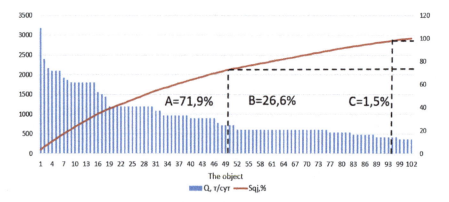

Fig. 1. ABC analysis of grain storage railway stations productivity (step 1)

As a result of ABC analysis, the grain storage railway stations were divided into groups: A – the largest and large grain storage railway stations, B – medium, C – small grain storage railway stations (step 1).

There is proposed to combine small grain storage railway stations of group C (due to low interest in them from the point of view of the organization of the transportation process) to merge with large groups A and B for reduce the sample size of the studied objects and the formation of grain clusters. The gravity model's modification was used for combine the grain storage railway stations. We use the fractional-power function of the gravity model for characterize the measure of "gravity" of grain storage railway stations (W_{ij}) [11]. The greater the value of the indicator, the more stable the transport links between the grain storage railway stations. Estimated "gravity" presented in Table 2. There are 5 positions in group C, as a result of ABC analysis.

Table 2. The magnitude of the gravitational bonds of grain storage railway stations (W_{ij}) for combining objects of group C (calculation of productivity).

Object (station)	Object 35	Object 12	Object 84	Object 95	Object 96	Object 68
Object 98	367.35	864	42.81	4.99	4.55	3.72
	Object 76	Object 90	Object 92			
Object 99	281.25	349,03	54.69			
	Object 12					
Object 100	495.87					
	Object 15	Object 29				
Object 101	803.31	426.03				
	Object 92	Object 95	Object 96	Object 12	Object 98	Object 68
Object 102	58.33	11.52	9.64	15.41	1.,87	5.14

Note: the stations with the highest indicator of "gravity" (W_{ij}) are italicized.

The gravitational connections analysis showed the most effective options for combining small grain storage railway stations (group C) with larger ones.

Object 90 moved from Medium to Large, Object 92 went up in its group, Object 12 and Object 15 went up in its group after combining group C grain storage railway stations with more powerful grain storage railway stations. Next, step 2 of the ABC analysis was performed, which provides for ranking with unification using a gravity model and obtained the groups A*, B*, C*.

The grain storage railway stations defined in group A* were selected (18 positions) as a result of the ABC analysis (step 2) in terms of "productivity", 10 grain storage railway stations of which were selected with the highest capacity indicators (Fig. 2).

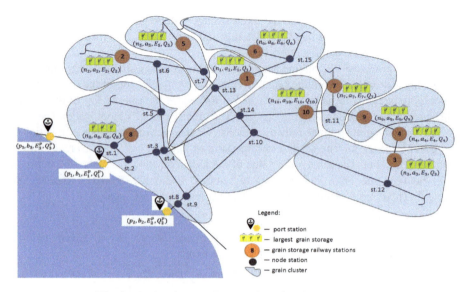

Fig. 2. Grain storage railway stations location at the loop

4 Two Criteria Optimization of the Cargo Transportation Process

If we consider the research carried out in this work in a general ideological sense, then they are in line with an egalitarian approach to the theory of welfare [12]. The implementation of the paradigm of cooperative decision making is presented in the form of an optimization problem with two objective functions. As the first criterion, the total transportation time is considered, and the second is the total time spent by rail routes at the railway loop.

We will consider m grain storage railway stations and n port stations. Let a_i be the number of routes to be sent from the i ($i = 1, 2, \ldots, m$) grain storage railway station, b_j – the number of routes that can be taken by the j ($j = 1, 2, \ldots, n$) port station. We will assume that all the quantity of grain planned for export from the grain storage railway station can be accepted by the port stations, that is,

$$\sum_{i=1}^{m} a_i \leq \sum_{j=1}^{n} b_j \tag{1}$$

The method given is uniformly applicable both in situations with a fulfilled balance condition and in the case of an open model [13]. That is, in the cases when instead of the inequality (1) the corresponding strict inequality holds. In this case, the total reception capacity of grain at port stations exceeds the total amount to be exported from grain storage railway stations.

Let in each transport plan x_{ij} signifies a number of routes that can be routed from the i grain storage railway station to the j port station (with a natural standard restriction $x_{ij} \geq 0$).

As feasible plan, we will consider transport plan (x_{ij}) which out to meet the following two conditions. First, all routes planned for removal at each of the grain storage railway station must be removed (but no more than that), that is, equality must be holds:

$$\sum_{j=1}^{n} x_{ij} = a_i \ (i = 1, 2, \ldots, m) \tag{2}$$

Secondly, the total number of routes from all grain storage railway station stations to each of the port stations should not exceed the number of routes that this port station can accept, that is, inequalities must be satisfied:

$$\sum_{i=1}^{m} x_{ij} \leq b_j \ (j = 1, 2, \ldots, n) \tag{3}$$

Expanding the standard formulation of the transportation problem, we proceed from the fact that in real situations the balance condition may not be satisfied [13].

We now turn to the temporary indicators of the transportation process. Let be t_{ij} the time the route was in the area between the i grain storage railway station and the j port station. The transportation time between all grain storage railway station and all port stations can be represented by a matrix (t_{ij}). The matrix, whose elements are the numerical values t_{ij} considered in this paper, is contained in Table 3.

Table 3. Distances and time of the route on the sections between grain storage railway stations and port stations.

Grain storage railway stations	Port stations					
	Novorossijsk		Tuapse		Taman	
	l_{ij}	t_{ij}	l_{ij}	t_{ij}	l_{ij}	t_{ij}
Beloglinskaya (1)	341	1.06	353	1.096	412	1.28
Starominskaya -Ejskaya (2)	276	0.86	327	1.02	331	1.03
Zelenokumsk (3)	598	1.86	496	154	669	2.08
Budennovsk (4)	634	1.97	546	1.7	705	2.19
Kry'lovskaya (5)	327	1.02	339	1.05	398	1.24
Celina (6)	460	1.43	472	1.47	531	1.65
Ipatovo (7)	536	1.66	548	1.7	607	1.89
Angelinskaya (8)	121	0.38	268	0.83	176	0.55
Blagodarnoe (9)	561	1.74	573	1.78	632	1.96
Spicevka (10)	452	1.4	464	1.44	523	1.62

Recall x_{ij} represents the number of routes that can be routed from any grain storage railway station to a port station. Let us now pay attention to the fact that the possible values x_{ij} are very insignificant compared to the number of all routes located at the railway loop at any given time. Thus, it is correct to assume that t_{ij} does not depend on x_{ij} values. Moreover, the numbers placed in Table 3 have been founded as average statistical values of the durations of the time spent by the routes in the respective sections, provided that many hundreds of cargo routes for various purposes are located on the railway loop.

So, in accordance with what was said at the beginning of the section, we will consider as the first objective function the next

$$z_1 = \max_{i,j}\{t_{ij}\,\text{sgn}\,x_{ij}\} \qquad (4)$$

where as usual

$$\text{sgn}\,x = \begin{cases} -1, & x<0, \\ 0, & x=0, \\ 1, & x>0. \end{cases} \qquad (5)$$

(Note that due to the non-negativity of the values under consideration, only the 2nd and 3rd lines in expression (5) are used.)

Thus, in the transport plan under consideration, z_1 is the time of the carriage for which it is the maximum. Since the value z_1 determines the time of implementation of the entire transportation plan, it is called the total transportation time. Within the framework of considering only one objective function (4) (which is nonlinear with respect to variables x_{ij}), the problem of integer mathematical programming $z_1 \to \min$ with constraints (2) and (3) can be considered. In this situation, the best is a plan in which the time of maximum transportation time does not exceed the time of maximum transportation time in any other plan.

Along with the objective function z_1 introduced above, we will consider the objective function of the form:

$$z_2 = \sum_{i=1}^{m}\sum_{j=1}^{n} t_{ij}\,x_{ij}. \qquad (6)$$

For each transport plan (x_{ij}), value z_2 represents the sum of all time spent on the railway loop by all routes that participate in the implementation of this plan.

Within the framework of considering only one objective function (6) (which is linear with respect to variables x_{ij}), the problem of integer mathematical programming $z_2 \to \min$ with constraints (3) and (4) can be considered. In this situation, a plan is optimal in which the sum of all time spent on the railway loop by all routes that participate in the implementation of this plan does not exceed the sum of all time spent on the railway loop by all routes that participate in the implementation of any other plan.

Next, we will consider an optimization problem with some set D of feasible transport plans and two objective functions z_1 and z_2. To each transportation plan

$(x_{ij}) \in D$, we assign a vector $\{z_1, z_2\}$, called a utility vector. Let us call the optimal transportation plan $(x'_{ij}) \in D$ with a utility vector $\{z'_1, z'_2\}$ for which there is no transportation plan such that the coordinates of its utility vector $\{z_1, z_2\}$ satisfy the conditions:

$$(z_1 \leq z'_1 \text{ and } z_2 < z'_2) \text{ or } (z_1 < z'_1 \text{ and } z_2 \leq z'_2). \tag{7}$$

5 Provision of Internal Stability for Cooperation of Participants of the Traffic Process

We give a brief analysis of the two indicators 1 and 2 of the transportation process introduced in the previous section.

Time z_1 is the direct indicator of the traffic plan (x_{ij}) by which (calculated in days) business interest of the carrier companies expresses. All efforts of each operator company are naturally aimed at delivering all the cargoes entrusted to it to the recipient as quickly as possible and without residue. That is, to implement such a transportation plan (x_{ij}), in which the corresponding time would be the minimal. That is, to implement such a transportation plan 1, in which the corresponding time would be the minimal.

Time z_2 is an indicator of the transportation plan (x_{ij}), which expresses (also calculated in days) the degree of operation of the track infrastructure of the loop when implementing this plan. (As is known, the owner of this infrastructure in Russia is the state, represented by Russian Railways.) It is important to pay attention to the fact that minimization of indicator z_2 is in full compliance with the socially oriented principles of green logistics. At present, as is known, the implementation of such principles is actively welcomed not only by government agencies, but also by the business community. Many advanced companies, demonstrating social responsibility, develop a strategy for their activities not only taking into account their own interests, but also taking into account the values declared by society as a whole.

In view of the above, it will be correct to note that the optimization task, stated in the previous section, is in the trend of corporate social responsibility, in which they try to achieve a certain balance between profitability, efficiency and social interests [14].

In the next section, for the problem we are considering of finding the best time transportation of grain cargoes, the corresponding implementation of an egalitarian approach in welfare theory will be presented [12]. As you know, one of the basic rules of this approach is expressed by the desire to equalize the individual utility of cooperative agents. In the mathematical model proposed in the previous section, it is natural for agents to consider an operator company that has taken on the responsibility for the transportation of goods, and Russian Railways, which represents the owner of the track infrastructure. The only controlled variable in this case is the utility, in the role of which is time, considered from the point of view of its possible minimization. It would like to note that the objective functions z_1 and z_2 considered in this case are not only not antagonistic, but (contrary to the terminology used in welfare theory) are not even conflicting with each other.

6 Implementation of Computational Procedures Using Software

For the two-criterion problem of mathematical programming stated in the previous section, an algorithm for finding the optimal distribution of cargo flows has been developed and implemented. At the same time, based on the operative situation, one can vary the set of permissible transportation plans. As software, Free Ware can be used, for example, Maxima.

In accordance with what has been stated in Sect. 4, the optimization algorithm is implemented by means of a conditional operator, in which the logical connective (7) is used. The target function z_1 is the total time of transportation (see (4)), and the objective function z_2 is the total time spent on the railway loop for all routes that make up the transportation (see (6)). From the expression (7) it can be seen that with the considered approach to the optimization process, the objective functions z_1 and z_2 are in the same position.

Let us proceed to the presentation of the results of computational procedures. When formulating conditions for transportation plans, we use expert estimates. We will assume that the port stations of Novorossiysk, Tuapse and Taman arrive no more than 25, 10 and 7 routes, respectively. From Fig. 2 and Table 3, it can be seen that Angelinskaya station is located much closer to Novorossiysk station than to the two other port stations under consideration. Therefore, it is natural to assume that at least two routes go from Angelinskaya station to Novorossiysk and at the same time not a single route goes to Taman. In addition, at least one route should be sent from each of the Ipatovo and Blagodarnoe stations to Novorossiysk, and not a single route should be sent to Taman. All routes from Spitsevka station should be sent to Novorossiysk.

Table 4. Feasible transportation plans.

№	Feasible transportation plan	Number routes arriving at port stations			z_1	z_2
		P_1	P_2	P_3		
1	0,0,3,0,0,3,0,1,1,0,2,0,0,2,0,0,3,0,1,1,0,2,1,0,2,0,0,3,0,0	8	10	7	2.08	33.09
.	....	.	.	.	.	.
9	0,0,3,0,0,3,0,1,1,0,2,0,2,0,0,3,0,0,2,0,0,3,0,0,2,0,0,3,0,0	15	3	7	2.08	32.42
10	0,0,3,0,0,3,0,2,0,0,2,0,0,1,1,0,3,0,1,1,0,3,0,0,1,1,0,3,0,0	8	10	7	1.78	32.33
11	0,0,3,0,0,3,0,2,0,0,2,0,0,1,1,0,3,0,1,1,0,3,0,0,2,0,0,3,0,0	9	9	7	1.74	32.29
.	....	.	.	.	.	.
65	2,1,0,3,0,0,0,2,0,0,2,0,1,1,0,3,0,0,2,0,0,3,0,0,2,0,0,3,0,0	19	6	0	1.74	30.78
66	2,1,0,3,0,0,0,2,0,0,2,0,2,0,0,3,0,0,2,0,0,3,0,0,2,0,0,3,0,0	20	5	0	1.74	30.75
67	3,0,0,3,0,0,0,2,0,0,2,0,1,1,0,3,0,0,2,0,0,3,0,0,2,0,0,3,0,0	20	5	0	1.74	30.74
68	3,0,0,3,0,0,0,2,0,0,2,0,2,0,0,3,0,0,2,0,0,3,0,0,2,0,0,3,0,0	21	4	0	1.74	30.71

In Table 4 is given 8 of the 68 feasible transportation plans that are consistently calculated by the software. For the optimal plan given in the last line, the total transportation time was equal to 1.74 days, and the total time spent on the railway loop for all the routes forming this plan was equal to 30.71 days.

We make a general observation of a methodological nature. In the proposed approach to the search for the optimal transportation plan, a significant role is given to the mathematical experiment, which is a heuristic tool in conducting research carried out in the framework of simulation modeling. Based on the operational situation and expert recommendations, it is possible to purposefully and flexibly manipulate the values of the numerical parameters characterizing transportation. The results below provide some insight into the varying possibilities of the approach when modeling freight traffic processes.

From the transportation plans in the last four lines of Table 4 show that as a result of their optimization, the distribution of routes by port stations takes the form: Novorossiysk – 21, Tuapse – 4 and Taman – 0. In this connection, we want to pay the attention to one important circumstance. The period from August to October (one of the peak for shipment of grain grown in the south-western Russian region) coincides with the period of peak transit transportation of grain grown in areas of the country remote from seaports. Therefore, during this period, it is advisable to at least partially unload the port station of Novorossiysk by directing some routes at Tuapse and Taman stations. So, using the results of expert assessments, we will additionally assume that no less than 10, but no more than 15 routes should go to the Novorossiysk station, not less than 6 and not more than 9 to the Tuapse station, and not less than 4 and not more than 11 routes to the Taman station.

Table 5. Feasible transportation plans.

№	Feasible transportation plan	Number routes arriving at port stations			z_1	z_2
		P1	P2	P3		
1	2,0,1,1,0,2,0,0,2,0,0,2,0,0,2,0,3,0,1,1,0,2,1,0,1,1,0,3,0,0	10	6	9	2.19	34.42
...	....	.	.	.	.	.
16	2,0,1,1,0,2,0,0,2,0,1,1,0,2,0,0,3,0,2,0,0,3,0,0,2,0,0,3,0,0	13	6	6	2.19	33.02
17	2,0,1,1,0,2,0,0,2,0,2,0,0,0,2,0,3,0,1,1,0,3,0,0,1,1,0,3,0,0	11	7	7	2.08	32.99
...	....	.	.	.	.	.
39	2,0,1,1,0,2,0,1,1,0,2,0,0,2,0,2,1,0,2,0,0,3,0,0,2,0,0,3,0,0	15	6	4	2.08	31.91
40	2,0,1,1,0,2,0,2,0,0,2,0,0,0,2,0,3,0,1,1,0,3,0,0,1,1,0,3,0,0	11	9	5	1.78	31.91
41	2,0,1,1,0,2,0,2,0,0,2,0,0,0,2,0,3,0,1,1,0,3,0,0,2,0,0,3,0,0	12	8	5	1.74	31.87
...	....	.	.	.	.	.
46	2,0,1,1,0,2,0,2,0,0,2,0,0,1,1,1,2,0,2,0,0,3,0,0,2,0,0,3,0,0	14	7	4	1.74	31.60
47	2,0,1,1,0,2,0,2,0,0,2,0,0,1,1,2,1,0,2,0,0,3,0,0,2,0,0,3,0,0	15	6	4	1.74	31.56
48	2,0,1,1,0,2,0,2,0,0,2,0,0,2,0,2,0,1,2,0,0,3,0,0,2,0,0,3,0,0	15	6	4	1.74	31.55

Table 5 shows 9 out of 48 feasible transportation plans sequentially calculated by the software.

Table 6. Optimal transportation plan.

Grain storage railway stations	Port stations		
	Novorossijsk	Tuapse	Taman
Beloglinskaya (1)	2	0	1
Starominskaya -Ejskaya (2)	1	0	2
Zelenokumsk (3)	0	2	0
Budennovsk (4)	0	2	0
Krylovskaya (5)	0	2	0
Celina (6)	2	0	1
Ipatovo (7)	2	0	0
Angelinskaya (8)	3	0	0
Blagodarnoe (9)	2	0	0
Spicevka (10)	3	0	0

For the considered two-criteria optimization task, the total time of the optimal transportation plan (see the last row of Table 5) was equal to 1.74 days, and the total time spent on the railway loop for all the components of this plan was 31.55 days. The vector of utilities with the corresponding coordinates satisfies the unanimity principle and is Pareto optimal [12]. Table 6 shows the number of routes that make up the optimal transportation plan, correlated with the port stations and grain storage railway stations.

The method of distributing export grain cargo flows presented in this article, according to the optimization criteria introduced, is of a general nature and can be used for both rail and motor transport.

7 Conclusion

A technique has been developed for a general approach to the optimization according to time criteria distribution of cargo flows in the regional transport and technological system. A mathematical model of the transportation process, which is a two-criterion problem of nonlinear integer programming and applicable to various types of transport has been formed.

As a specific object of application of the proposed optimization method, the distribution of grain cargo flows in the regional railway transportation and technological system is considered. A technique has been developed for selecting clusters of bulk cargoes on a given network, taking into account regional infrastructure indicators and economic and geographical parameters of cargo accumulation. The algorithm has been implemented and for a number of cases the procedure of finding the optimal distribution of cargo flows to the port stations has been brought to numerical results.

The developed method can obtain applications in formulating operational recommendations when adjusting the distribution of cargo flows in the region.

Acknowledgements. The research is conducted with a support of the Russian Foundation for Basic Research (#17-20-04236 ofi m RJD).

References

1. Zhang, Y.Z., Wang, J.Q., Hu, Z.A.: Optimization model of transportation product selection for railway express freight. J. Eng. Sci. Technol. Rev. **9**(5), 104–110 (2016)
2. Knoop, V., Hoogendoorn, S.: An area-aggregated dynamic traffic simulation model. Eur. J. Transp. Infrastruct. Res. **15**(2), 226–242 (2015)
3. Wang, X., Meng, Q., Miao, L.: Delimiting port hinterlands based on intermodal network flows: model and algorithm. Transp. Res. Part E Logist. Transp. Rev. **88**, 32–51 (2016)
4. Gooley, T.B.: The Geography of Logistics. Logistics Management and Distribution Report January, pp. 63–65 (1998)
5. Harrison, R.: International Trade, Transportation Corridors, and Inland Ports: Opportunities for Canada. Pacific-Asia Gateway and Corridor Research Consortium, pp. 1–13. http://www.gatewaycodor.com/roundconfpapers/documents/Harrison_Robert_Winnipeg.pdf
6. Dinu, O., Burciu, S., Oprea, C., Ilie, A.: Inland waterway ports nodal attraction indices relevant in development strategies on regional level. IOP Conference Series: Materials Science and Engineering (2016)
7. Combes, F., Tavasszy, L.A.: Inventory theory, mode choice and network structure in freight transport. Eur. J. Transp. Infrastruct. Res. **16**(1), 38–52 (2016)
8. Chislov, O., Bogachev, V., Zadorozhniy, V., Bogachev, T.: Economic-geographical method delimiting wagon flows in the region considered: model and algorithm. Transp. Probl. **13**(2), 39–48 (2018)
9. Hyland, M., Mahmassani, H., Mjahed, L.: Analytical models of rail transportation service in the grain supply chain: deconstructing the operational and economic advantages of shuttle train service. Transp. Res. Part E Logist. Transp. Rev. **93**, 294–315 (2016)
10. Karabatsos, G., Leisen, F.: An approximate likelihood perspective on ABC methods. Statist. Surv. **12**, 66–104 (2018)
11. Chislov, O.N., Ljuts, V.L.: Modified gravity method in placement the distribution of port terminals rail transport technological systems. Eng. J. Don **23, 4–2** (23), 82 (2012)
12. Moulin, H.: Axioms of Cooperative Decision Making (Econometric Society Monographs). Cambridge University Press, New York, Cambridge (1989)
13. Vinogradov, I.M.: Mathematical Encyclopedia. Soviet Encyclopedia, vol. 5, pp. 419–421 (1982)
14. Mindur, L., Hajdul, M.: The concept of organizing transport and logistics processes, taking into account the economic, social and environmental aspects. Transp. Probl. **8**(4), 121–128 (2013)

Tools Supporting the Reduction of Negative Environmental Impact in Transport

Strategic Planning of the Development of Trolleybus Transportation Within the Cities of Poland

Marcin Wołek[(✉)] and Katarzyna Hebel

Chair of Transportation Market, Faculty of Economics, University of Gdańsk,
Gdańsk, Poland
{marcin.wolek,katarzyna.hebel}@ug.edu.pl

Abstract. In recent years, trolleybus transport is undergoing technological transformation from catenary-based electric public transport to flexible electromobility platform based on 'hybrid-trolleybuses' with batteries. The article presents an analysis of state-of-the-art of trolleybus transport in Poland, including three urban areas. The research was conducted within TROLLEY 2.0 project. An analysis of quantitative data has been carried out in connection with the usage and economics of selected trolleybus operators. Additionally, an in-depth analysis of strategic documents adopted by the local authorities of those cities of Poland which have an operating trolleybus network was conducted. It leads to a conclusion that the quantitative and qualitative development of trolleybus transportation in Poland took place especially after EU accession.

Keywords: Trolleybus transport · Electromobility · Strategic planning · Transport economics

1 Introduction and Methodology

Urban transport is a naturally a suitable area for the implementation of electromobility solutions [1]. Recent years have seen the dynamic development of hybrid and electric buses. This article discusses a means of transportation that has been running on electric power since 1882. The quantitative development of trolleybus transportation reached its peak in the 50's, with technical development allowing for a decrease in demand for vehicles depending on catenary thanks to diesel engine buses. An increase in the importance of environmental issues, as well as the progress of technology connected with power storage, made it possible for a trolleybus to be redefined as an in-motion-charging electric bus [2] or "hybrid trolleybus". The potential for trolleybuses to travel outside catenary was pointed out in academic articles nearly a decade ago [3] as well as in research projects partially funded by the EU (CIVITAS DYN@MO and ELIPTIC) [4, 5]. Thanks to the growing number of trolleybuses equipped with modern traction batteries the distance travelled by these vehicles outside the catenary continues to grow.

Among the advantages of trolleybus transportation discussed in other works is its eco-friendliness [6] although this is said also about other vehicles charged by various technologies alternative to diesel powertrain ("in overall, the WTW evaluation shows that BEB has great potential to reduce GHG emissions" [7, 8]. The higher emissions at the production stage of BEV are compensated over its lifetime [9].

It appears there is no need to further discuss the aspects of the superior energy efficiency of electric vehicles. Energy consumption analysis (Well-to-Wheel) shows that a reduction of 50% per kilometre is possible in comparison to diesel buses [10], but is strongly dependant on the share of electricity produced from renewable sources [11].

Thanks to the in-motion-charging technology hybrid trolleybuses are equipped with a battery of a respectably lower capacity than that of classic battery-powered electric buses (BEB) [12]. The tendency to ensure a high level of independence of a battery-powered bus makes for the necessity to equip it with a high capacity battery [13], This inflates the cost of such a vehicle as well as increasing its weight. Consequences of reduced operational performance are the scheduling of battery buses, the fleet composition, and the optimization of charging infrastructure (spatial deployment, power and charging time) [14].

The question of battery life seems to be appropriate when discussing battery-powered electric buses (BEB) since it is the battery that highly influences the cost of such vehicles. Though the life-span of a bus itself can extend past 20 years, the life-span of a battery is much shorter. LCC modelling points to the necessity to increase additional costs connected with the exchange of batteries in BEBs [15].

It emerges from the readings of subject matter literature that battery-equipped trolleybuses are a reliable service provider in urban transportation.

An analysis of quantitative data has been carried out in connection with the usage and economics of selected trolleybus operators in Poland. Additionally, an in-depth analysis of strategic documents adopted by the local authorities of those cities of Poland which have an operating trolleybus network (Gdynia, Lublin and Tychy).

2 Economic and Operational Determinants of Trolleybus Transportation in Poland in 2017

The share of urban transport vehicles with alternative drives in Poland in 2017 was relatively low [16] and amounted to 6% (including trolleybuses). 219 trolleybuses amounted to 1.8% of the total number of on-road vehicles (exclusive of rail-traction vehicles) which were in operation as part of urban transportation network in the cities of Poland.

In Poland there are currently 3 cities with an operational trolleybus network, namely Lublin, Gdynia (inclusive of Sopot), and Tychy. Neither of those cities has an operational tram network. However, in 2017 Lublin launched service operated by electric buses but scale of operations is very low up to now (Table 1).

Table 1. Types of tractions in the urban transportation networks of Gdynia, Lublin, and Tychy (source: own elaboration)

	Trolleybus	Electric bus	Hybrid bus	CNG bus	Diesel bus
Gdynia and Sopot	X	–	–	X	X
Lublin	X	X[a]	–	–	X
Tychy	X	–	–	X	X

[a]tests

Tychy, Lublin, and Gdynia are currently undergoing procedures connected with the purchase of electric buses. Additionally, in Tychy there is an open tender for the purchase of more CNG buses.

Table 2 presents an overview of the basic data on supply of trolleybus transport. Although MPK Lublin owns the largest amount of trolleybuses, it is in Gdynia and Sopot where trolleybuses were are in operation the most.

Table 2. Basic data on trolleybus transportation in Poland in 2017 (source: own elaboration)

	Gdynia and Sopot	Lublin	Tychy
Established in	1943	1953	1982
Number of vehicles	90	108	21
Share of articulated vehicles [%]	0	11%	0
Number of connections [exclusive of end-of-route, seasonal and tourist]	11	11	6
Operational volume [thousands vehicle-kilometres]	5 199.1	4 981.4	1 309.4
Number of routes with trolleybuses travelling independently of catenary	2	7[a]	0

[a]based on regular timetable, currently the number of lines partially operating without catenary is 11 due to road reconstruction

In total, trolleybuses in service within these cities completed nearly 11.5 m vehicle-kilometres, with the majority of this distance (88%) travelled in Gdynia and Sopot as well as Lublin (Fig. 1).

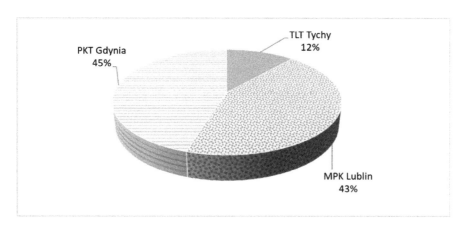

Fig. 1. Division of trolleybus transportation operational work in Poland in 2017 between operators (source: own study on the basis of internal data of operators, and strategic documents of Gdynia, Lublin, and Tychy)

Joining the European Union kick-started the process of change within urban public transport in Poland in the above cities. The increase in the importance of the creation of high-quality life standards inclusive of eco-awareness, was an impulse for the development of low-emission urban transportation. This development has been (and still is) both qualitative and quantitative. Quantitative development is expressed in the growth in demand for trolleybus transport as expressed in operational work (vehicle-kilometres). In the years 2011–2017 the supply of trolleybus transport in Poland grew by 29% (Fig. 2). Lublin manifested the highest dynamics of growth - supply of trolleybus transport in this city during the period mentioned grew by 84%. In Gdynia and Sopot this amounted to 5%, and in Tychy to 3%.

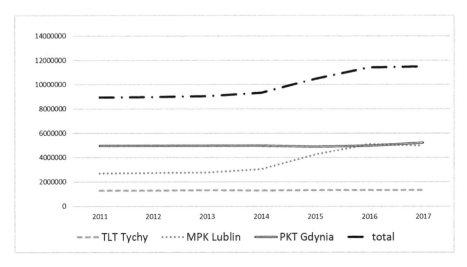

Fig. 2. Supply of trolleybus transport in Poland in the years 2011-2017, according to operator [veh-kms] (source: own study on the basis of internal data of operators and strategic documents of Gdynia, Lublin and Tychy)

A measurable indicator of qualitative growth in trolleybus service supply can be seen in the average age of rolling stock. This decreased in comparison with 2005 (one year after EU accession) amounting to 9 years in 2012 and 5 years in 2017 in case of operator MPK Lublin (Fig. 3).

The dynamics of change in this respect can be illustrated by the scale of supply of new trolleybus rolling stock. In 2003 all three trolleybus operators purchased 1 new vehicle, whilst in 2018 this number grew to 18 (15 purchased by MPK Lublin and 3 by PKT Gdynia). Purchases of new trolleybus vehicles are largely funded by non-repayable EU funds. An analysis of one of the operators (PKT Gdynia) shows that in the years 2001–2016 18 vehicles were purchased solely from own assets, whilst 2 vehicles were purchased as part of CIVITAS DYN@MO project with 25% funding. 38 trolleybuses were purchased with the help of ERDF funds covering between 50 and 75% of the total price (Fig. 4). In the analysed period there was another way in which the quality of the rolling stock was affected - the conversion of diesel buses into trolleybuses carried out by the operator themselves (also referred to as fleet renewal). In

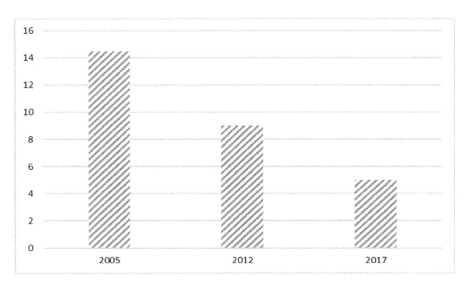

Fig. 3. The average age of trolleybus rolling stock on the example of MPK Lublin in 2005, 2012 and 2017 (source: own study on the basis of internal data of MPK Lublin)

initial phase, conversion costs amounted to ca. 26–37% of the total price of a brand new vehicle [17], although the conversion process was not very advanced due to lower complexity of diesel buses. It allowed for a substantial increase the number of low floor vehicles owned in a short period of time and constituted a transitional stage in the complex renewal strategy of trolleybus rolling stock in Gdynia. The following vehicles were converted: Mercedes O405 N (27 vehicles), Mercedes Citaro (2 vehicles), and Solaris Urbino (3 vehicles).

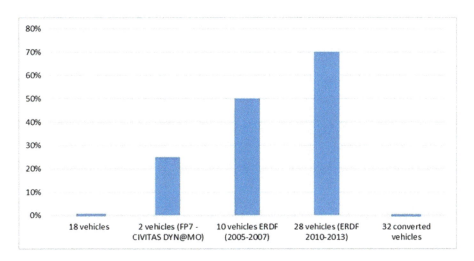

Fig. 4. Level of grant support in financing trolleybus fleet in Gdynia 2001–2016 (source: own study on the basis of internal data of PKT Gdynia).

Each of the operators of trolleybus transport in Poland provides services as a limited liability company owned by local authorities. Operators in Gdynia and Tychy use trolleybuses exclusively, whilst MPK Lublin also uses diesel and electric buses.

The cost structure of TLT Tychy provides evidence of the increasing role of personal costs. In 2017 43% of general costs incurred were connected with salaries and similar expenses (Fig. 5).

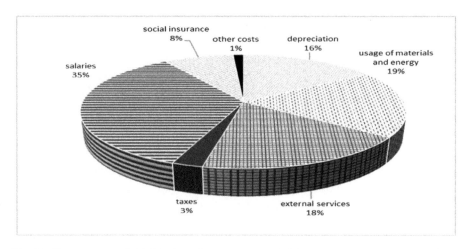

Fig. 5. Cost structure of TLT Tychy trolleybus operator in 2017 (source: own study on the basis of internal data of TLT Tychy)

Remuneration from public transport authorities (based on in-house gross contract) per vehicle-kilometre differed significantly and the lowest value was achieved in Tychy (Table 3).

Table 3. Remuneration from public transport authorities per vehicle-kilometre for services provided by trolleybuses in 2017 in Poland (Source: own study on the basis of internal data of TLT Tychy, PKT Gdynia and MPK Lublin)

Operator	Value per vehicle-kilometre [PLN]
TLT Tychy	7.12
PKT Gdynia	8.78
MPK Lublin	8.98

The cost of vehicles is a major obstacle in the expansion of trolleybus network (Table 4). The cost of 12 m hybrid trolleybus (equipped with a battery) is twice as high as the cost of a standard diesel bus that complies with the EURO6 norm. It has to be, however, emphasised that trolleybuses have a longer life cycle of over 20 years.

Table 4. The cost of the purchase of buses in Poland during the years 2017–2018 (Source: own study on the basis of internal data of operators)

Type of vehicle	Operator	City [year of purchase]	Type of vehicle	Number of vehicles purchased	Unit value [net, without VAT in [EUR]
12 m diesel bus (EURO6)	MPK Wloclawek	Wloclawek [2018]	MAN Lion City A37	4	221 000
12 m CNG bus	MPK Tarnow	Tarnow [2017]	Scania Citywide 12LF CNG	21	257 125
12 m hybrid bus	KM Plock	Plock [2018]	Solaris Urbino H12	17	425 000
12 m trolleybus with a battery	TLT Tychy	Tychy [2018]	Solaris Trollino 12	3	466 250
12 m electric bus	MPK Krakow	Krakow [2017]	Solaris New Urbino 12 Electric	17	502 198

[1 EUR = 4 PLN]

3 Analysis of Clauses Regarding Trolleybus Transport in the Strategic Documents of Gdynia, Lublin and Tychy

Public transport is part of the responsibility of the local authorities of Polish cities. Amongst nearly 2500 self-governing communes there are 66 urban districts where the authority of the commune and the districts are intertwined within in the borders of a city. Although the largest proportion of expenditure is generated by the education sector [18], it is transportation that is the second largest source of expense. Below is an analysis of clauses included in strategic documents of the highest importance in use by the cities with operational trolleybus networks. It is inclusive of acts passed by the City Council which are in part concerned with transportation issues.

Some of the documents are obligatory for cities (i.e. Low-Emission Economy Plan or of Plan of Sustainable Development of Public Transport).

In Gdynia, plans for development include two scenarios for trolleybuses. The first being to maintain the current level of operations. In the 2nd, trolleybus transport shall be developed in case of spatial scope and annual volume of operations.

In Lublin plans for trolleybus development included extension of catenary network and introduction of hybrid trolleybuses.

In Tychy, trolleybus transport was highly prioritized, mainly to its environmental advantages.

The possibility of recharging the battery while operating within the network will give trolleybuses an important advantage over ordinary electric buses that need stationary form of charging (opportunity, overnight or combination of both) and it is clearly stated in strategic documents of Lublin and Gdynia.

In all cities ecological factor is very important, especially focus on decreasing the negative impact of transport on the environment (planned increase of the share of trolleybuses in the public transport market) (Tables 5, 6 and 7).

Table 5. Strategic documents of Gdynia city in regards to trolleybus transport (Source: own study).

Document	Year	Type	Selected citations
Transport Policy of Gdynia City	1998	City Council Act	The following is planned in order to increase the quality of public transport: • continue the process of upgrading bus and trolleybus rolling stock, also in order to decrease negative environmental impact • continue upgrading catenary • provide proper standards for PRM persons
Strategy of Gdynia Development	2003	City Council Act	Development of trolleybus transport as environmentally friendly and co-creating unique city image
Integrated Plan for Public Transport Development in Gdynia 2004–2013	2004	City Council Act	Development of trolleybus transport as environmentally friendly and co-creating unique city image via the extension of the trolleybus network and the building of a new trolleybus depot Further upgrading of rolling stock
General Spatial Master Plan	2008	City Council Act	Introduction of new trolleybus lines into newly developed urban areas Development of a traffic management system for bus and trolleybus priority Development of trolleybus network
Plan for Sustainable Public Transport in Gdynia 2008–2015	2009	"internal", elaborated within BUSTRIP project	Development of a sustainable, integrated urban transport system as a means of creating the potential for safe, friendly and environmentally-friendly travel Direction 4: Development of clean and friendly means of transport
Covenant of Mayors	2011	City Council Act	Development and execution of actions focused on sustainable energy policy
Update of Plan of Sustainable Development of Public Transport for Gdynia and Other Communes 2016–2025	2016	City Council Act	The change of role of the trolleybus in Gdynia's public transport service would allow for the reconstruction of the bus route system, by limiting the number of direct bus connections
Sustainable Urban Mobility Plan	2016	City Council Act	Development of competitive public transport Increase of share of low-emission vehicles
Strategy of Gdynia Development: Gdynia 2030	2017	City Council Act	Further development of pro-ecological public transport

Table 6. Analysis of strategic documents concerning trolleybus transportation: Lublin (Source: own study).

Document	Year	Type	Decisions made [selected]
Transport Policy of Lublin	1997	City Council Act	Development of catenary as a priority investment and new routes connecting certain areas of the city with the built-up areas of the city centre Environmental advantage of trolleybuses
Development Strategy for Lublin	2008	City Council Act	Amongst the projects there is a mention of "the improvement in quality of bus and trolleybus connections inclusive of the implementation of new green technologies"
The plan of sustainable development of public transport for the commune of Lublin and neighbouring communes working in cooperation with Lublin in the scope of management of public transport	2012	City Council Act	Development of catenary (trolleybuses) as part of the aim to "decrease the negative effects of transport on the environment" The principles of transport policies included: "the treatment of the development of catenary as a priority investment in public transport" The document points to the possibility of substituting buses with trolleybuses on some existing routes
Development Strategy 2013–2020 for Lublin	2013	City Council Act	A purchase of with batteries is anticipated
Low-emission Economy Plan	2015	City Council Act	The priorities includes "the development of sustainable multi-modal urban mobility and low-emission transportation" achieved by means of "the development and broader utilization of public transport inclusive of the purchasing of low-emission rolling stock, electric or hybrid rolling-stock, and the development of catenary"
Sustainable Mobility Plan for Lublin	2015	City Council Act	Priority for public transport vehicles highlighted (bus lanes and traffic light priority) especially within the main transport corridors of the city Promotion of green travel options (public transport, including of electric trolleybuses, cycling)

Table 7. Analysis of strategic documents concerning trolleybuses: Tychy (Source: own study).

Document	Year	Type	Decisions made [selected]
Development Strategy 2000 for Tychy	1996	City Council Act	The development of the trolleybus network was counted amongst the tasks of the highest priority
Sustainable development plan for public transport	2013	City Council Act	One of the aims focuses on essential reduction of the negative effects that transportation has on the environment and to implement solutions and technologies which decrease noise and emissions. "This can be achieved by such means as the development of trolleybus transportation" Aim: development of trolleybus transportation for the protection of the environment
Development Strategy 2020+ for Tychy	2014	City Council Act	Strategic aims include "a low-emission city becoming a leader in sustainable development in regional connections". This is to be achieved by such means as public transport (bus-trolleybus-rail) of such quality that it will bear direct influence on the efficient functioning of the city. It is anticipated that "gradual modernisations to the trolleybus network shall be made"
Low-emission Economy Plan for Tychy	2015	City Council Act	The purchasing of modern trolleybus rolling stock and construction of photovoltaic power station (solar park) to meet power demands of the trolleybus traction

4 Conclusions and Discussion

The analysis allows for the following conclusions:

- Trolleybus transport in Poland is operational in only 3 urban areas, including 4 cities, however, there is no threat to its market position in neither of the locations. This has been proved by the analysis of operational data provided in this article.
- The analysis of strategic documents of Gdynia, Lublin, and Tychy leads to the conclusion that the ecological factor of trolleybus transportation is significant to the authorities of all three cities, despite the consciousness of an unfavourable energy mix in Poland.
- Special attention is paid to the purchasing of rolling stock that allows for catenary-independent travel (equipped with traction batteries). This was proved by the decisions made by trolleybus operators in respect of purchases made (PKT Gdynia, MPK Lublin, and TLT Tychy) and modernisation (PKT Gdynia). The share of

hybrid trolleybuses (equipped with a battery allowing for catenary-independent travel) in the rolling stock cross-section of all three operators is growing steadily.
- The newest developmental trend within trolleybus systems can be seen in the plans of construction of photovoltaic pharm (Gdynia, Tychy). Solar power will be fed directory to the traction network which will decrease the emission generated by this form of transportation.
- It has to be noted that trolleybus systems which are perceived as environmentally friendly in the future are to be supported by electric buses. The cities analysed are already planning to purchase electric buses supported with opportunity charging technology (Lublin and Gdynia).
- The strive to lower emissions generated by urban transportation is also reflected in the purchasing of CNG buses (Gdynia and Tychy).

It has been, therefore, confirmed that the quantitative and qualitative development of trolleybus transportation in Poland is indeed taking place. This would not have been possible without the prior strategic aims included in strategic documents of the cities included in this analysis. These assumptions allowed for the allocation of assets which made it possible to apply for external funding.

The Authors would like to emphasise that it is indispensable to include local public transport in strategic planning and to promote bold, modern technical and technological solutions which are economically efficient. The development of trolleybus transportation within cities in which this mode of transport is already in operation can be perceived as such.

Acknowledgements. The article has been prepared within project TROLLEY 2.0 – Trolleybus Systems for Smart Cities. ERA-NET-COFUND in the framework of Horizon 2020 (Electric Mobility Europe).

References

1. He, Y., Song, Z., Liu, Z.: Fast-charging station deployment for electric bus systems considering electricity demand charges. Transportation Research Board 97th Annual Meeting, 48, pp. 1–31 (2018)
2. Bartłomiejczyk, M.: Praktyczna aplikacja In: Motion Charging w Gdyni: trolejbusy w obsłudze linii autobusowych [Practical application of In Motion Charging in Gdynia: trolleybuses in servicing bus lines] Autobusy : Technika, Eksploatacja, Systemy Transportowe, 7–8(3092), 58–64 (2016). http://yadda.icm.edu.pl/baztech/element/bwmeta1.element.baztech-8c48629c-9cc8-41f1-949c-38a9f9a387f2/c/BartlomiejczykTTS78.pdf. (in Polish)
3. Kühne, R.: Electric buses - an energy efficient urban transportation means. Energy **35**(12), 4510–4513 (2010). https://doi.org/10.1016/j.energy.2010.09.055
4. CIVITAS DYN@MO Consortium: DYN@MO | CIVITAS (2016). EU website http://civitas.eu/content/dynmo. Accessed 18 May 2018
5. ELIPTIC Project Consortium: ELIPTIC project (2018). http://www.eliptic-project.eu/. Accessed 23 Jan 2018

6. Corazza, M.V., Guida, U., Musso, A., Tozzi, M.: A European vision for more environmentally friendly buses. Transp. Res. Part D Transp. Environ. **45**, 48–63 (2016). https://doi.org/10.1016/j.trd.2015.04.001
7. Alessandrini, A., Cignini, F., Ortenzi, F., Pede, G., Stam, D.: Advantages of retrofitting old electric buses and minibuses. Energy Procedia **126**, 995–1002 (2017). https://doi.org/10.1016/j.egypro.2017.08.260
8. Eng, L., Ling, H., Yoke, S., Mun, L.: Scenario-based electric bus operation: a case study of Putrajaya. Int. J. Transp. Sci. Technol. **7**(1), 10–25 (2018). https://doi.org/10.1016/j.ijtst.2017.09.002
9. Rupp, M., Handschuh, N., Rieke, C., Kuperjans, I.: Contribution of country-specific electricity mix and charging time to environmental impact of battery electric vehicles: a case study of electric buses in Germany. Appl. Energy **237**, 618–634 (2019). https://doi.org/10.1016/j.apenergy.2019.01.059
10. Andwari, M.A., Pesiridis, A., Rajoo, S., Martinez-Botas, R., Esfahanian, V.: A review of battery electric vehicle technology and readiness levels. Renew. Sustain. Energy Rev. **78**, 414–430 (2017). https://doi.org/10.1016/j.rser.2017.03.138
11. Wolański, M., Wołek, M., Jagiełło, A.: Jak analizować efektywność finansową i ekonomiczną napędów alternatywnych? [How to analyse financial and economic efficiency of alternative powertrains?]. Biuletyn Komunikacji Miejskiej **148**, 6–12 (2018). (in Polish)
12. Gao, Z., Lin, Z., LaClair, T.J., Liu, C., Li, J.M., Birky, A.K., Ward, J.: Battery capacity and recharging needs for electric buses in city transit service. Energy **122**, 588–600 (2017). https://doi.org/10.1016/j.energy.2017.01.101
13. Lajunen, A.: Lifecycle costs and charging requirements of electric buses with different charging methods. J. Clean. Prod. **172**, 56–67 (2018). https://doi.org/10.1016/j.jclepro.2017.10.066
14. Rogge, M., van der Hurk, E., Larsen, A., Sauer, D.U.: Electric bus fleet size and mix problem with optimization of charging infrastructure. Appl. Energy **211**, 282–295 (2018). https://ac-els-cdn-com.libproxy.mit.edu/S0306261917316355/1-s2.0-S0306261917316355-main.pdf?_tid=1c442de6-eb52-11e7-8cba-00000aab0f27&acdnat=1514412535_238aaafb79cb050308d5650c3db2a07f
15. Nurhadi, L., Borén, S., Ny, H.: A sensitivity analysis of total cost of ownership for electric public bus transport systems in swedish medium sized cities. Transp. Res. Procedia **3**, 818–827 (2014). https://doi.org/10.1016/j.trpro.2014.10.058
16. Dyr, T.: Costs and benefits of using buses fuelled by natural gas in public transport. J. Clean. Prod. **225**, 1134–1146 (2019)
17. Wyszomirski, O., Połom, M., Bartłomiejczyk, M., Dombrowski, J.: Konwersja autobusu z silnikiem diesla na trolejbus. [Conversion of diesel bus into trolleybus]. Gmina Miasta Gdyni, Gdynia, pp. 1–84 (2012). (in Polish)
18. Wołek, M.: Mobilność sektora oświaty w samorządzie lokalnym w Polsce [Mobility of the education sector in local government in Poland] Zeszyty Naukowe Wydziału Ekonomicznego Uniwersytetu Gdańskiego. Transport i Logistyka (75), 137–147 (2017). (in Polish)

Ways to Improve Sustainability of the City Transport System in the Municipal Gas-Engine Vehicles' Fleet Growth

Irina Makarova[✉], Ksenia Shubenkova, Larisa Gabsalikhova, Gulnaz Sadygova, and Eduard Mukhametdinov

Kazan Federal University, Naberezhnye Chelny, Russian Federation
ksenia.shubenkova@gmail.com, kamivm@mail.ru,
funte@mail.ru, sadygova_1988@mail.ru,
muhametdinoval@mail.ru

Abstract. Natural gas today is of increasing interest, as it allows to reduce harmful emissions into the atmosphere. The environmental impact of large parks due to large annual mileage is higher than personal vehicles, however, such a fleet is easier to manage through a single control center. Vehicles of waste collection on gas engine fuel will reduce harmful emissions into the atmosphere and the noise level in the morning from the vehicles. Studies on the safety of vehicles on compressed natural gas have confirmed the high level of this vehicles, however, the issues of its reliability remain relevant. The proposed method for predicting potential failures and service planning, as well as forecasting operating conditions, will allow to take into account the prospects for expanding the gas engines vehicles' fleet and ways to reduce the burden on the environment.

Keywords: Gas-engine garbage trucks · Environmental safety · Municipal vehicle · FMEA

1 Introduction

Activities evaluation and analysis manufacturers of modern cargo vehicles and buses shows that innovative processes are aimed at limiting energy consumption and reducing harmful emissions in the exhaust gases of engines, as well as at increasing the equipment for active and passive safety efficiency. The motorization negative effects, expressed mainly in environmental pollution by harmful products contained in exhaust gases, were recognized as one of the main critical factors for the society future development. The growing population urbanization, accompanied by an increase in the megalopolises number, compels us to think about the need for preferential use of public transport and municipal equipment vehicles that meets environmental requirements.

Alcohol, liquefied petroleum gas, compressed and liquefied natural gas, hydrogen and electricity can be considered as alternative fuel types that can replace petroleum fuel. Gas engine fuel is cheaper than diesel, so it is not surprising that many companies seek to convert municipal equipment vehicles to gas if it is convenient to organize their refuelling. Of course, gas fuel has its drawbacks. The weight of gas cylinders somewhat reduces the payload capacity of the waste truck. In addition, the gas engine torque is lower than that of a diesel engine, and fuel consumption increases. But environmental characteristic meets all standards. Municipal equipment vehicles have a constant movement route and consumes a large fuel amount daily, therefore it is one of the priority segments for the gas-engine industry development. Improving the safety and sustainability of the city's transport system is possible by gradually updating the fleet of communal equipment vehicles with its gas counterparts.

2 Problem Condition: Research of Ways to Increase the Transport's Economy and Environmental Friendliness

Given the growing demand for energy consumption and the various threats posed by air pollution, work is underway to reduce energy use. The research's authors indicate different ways to reduce energy consumption and, consequently, emissions in the transport sector. These works can be divided into two areas:

- new technical and technological solutions aimed at improving the fuel quality, switching to its low carbon types, changing the way of obtaining energy, fundamentally new types of power plants, as well as the use of new materials and design solutions to reduce vehicle weight;
- organizational and management decisions aimed at improving the vehicle operation's efficiency, including by optimizing the transportation process, as well as organizing high-quality and timely service and repair of vehicles, preventing failures, increasing operational reliability and safety.

2.1 Prospects for Reducing Emissions from Alternative Fuels

Vehicle emissions are one of the most serious air pollution causes in large cities. So, in Moscow and other large Russian cities, the automobile emissions proportion is more than 90% of the total pollutants emissions into the atmosphere. In cities with more developed industry, the contribution share of automobile exhaust gases is slightly less (about 80–90%). In Russia as a whole, motor vehicle emissions to atmosphere account for 42% of their total number [1].

The objective prerequisites for the interest growth in the use of gas as a motor fuel [2] in recent years are higher energy and environmental indicators compared to petroleum fuels. Of all the widely used motor fuels and technologies, natural gas provides the safest emissions of exhaust gases, has a significant impact on lubricating oils (30–40%). Thus, the vehicles transfer from gasoline to gas reduces on average five times the harmful emissions and the noise impact by half. In addition, gas does not contain the main pollutant of gasoline - sulphur, so even the most refined gasoline "Euro-5" cannot be compared at clean combustion with gas fuel. Another important factor is the more stable price for gas and its higher efficiency: at a lower price than diesel fuel, energy is almost the same – 0.95: 1.

Urban air pollution has become a major problem in China. Thus, in the densely populated and developed region of Beijing-Tianjin-Hebei (BTH), vehicle emissions are the main pollution sources [3]. The article authors on the emission inventories basis have established four policy scenarios to simulate changes in emissions. The simulation results showed that replacing old vehicles with new ones and improving the quality of the oil can reduce emissions. The article authors [4] indicate that vehicle emissions have become one of the key factors affecting urban air quality and climate change in the Pearl River Delta region. Therefore, analyzing different scenarios for the transport system development, the authors believe that the government should adopt long-term abatement strategies, such as updating emission standards, fuel quality and bus priority strategies, long-term pollution struggle.

Investigating the contributions to air pollution of gasoline vehicles (GV) and diesel vehicles (DV), heavy diesel vehicles (HDDV) and non-HDDV, the article authors [5] conclude that the contribution of HDDV to vehicle fleet emissions, calculated by EFs from China The "Guidelines for Inventory of Emissions on Roads" was underestimated. The article authors [6] claim that their research results showed a significant underestimation of LDGV emission factors in the EI guide in China. The main reason for the high emissions of passenger cars with a gasoline engine (LDGV) according to the article authors research results are catalytic converters' malfunctions. Strengthening the vehicles supervision with high emissions and enhancing the tests for the conformity of vehicles which are in exploitation important measures that will contribute to improving the environmental situation in China. Life cycle analysis (LCA) is performed using a combination of real-time speed data on fuel consumption for diesel/liquefied natural gas (LNG) heavy-duty vehicles (HDV) in China [7]. The authors conclude that direct energy consumption and life cycle energy use for HDV on LNG is about 7.4% and 6.2% higher than comparable diesel HDV, while about 8.0% is estimated to reduce in life cycle GHG emissions if diesel HDV is replaced with HDV LNG in China.

The goal of the study [8] was to compare the life cycle in terms of greenhouse gas (GHG) emissions of diesel and liquefied natural gas (LNG), used as fuel for heavy vehicles on the European market (EU-15). The life cycle analysis of these fuels was carried out in accordance with the methodology based on the ISO 14040: 2006 [9] and ISO 14044: 2006 [10] standards for the "environmental management" sector. The authors focused on the European scenario and on heavy vehicles, given their contribution to global GHG emissions. In particular, one of the issues was an assessment of the possible use of small liquefaction plants installed at service stations, since this aspect is associated not only with environmental considerations, but also affects the actual distribution of this fuel, which is limited by the lack of a local distribution network.

Alternative fuels are an attractive option for reducing emissions from internal combustion engines. Scientific studies [11–13] are devoted to the possibilities of using various alternative fuels types: biodiesel (vegetable oils), methanol, ethanol, hydrogen, biomass (organic material), biogas, natural gas, dimethyl ether, etc.

The document [14] discusses key environmental, technical and socio-economic aspects of the LNG deployment as an alternative fuel for road freight transport. Since the EU encourages the use of less polluting alternative fuels, while ensuring the supply safety and optimal energy storage, it is necessary to explore the prospects for the introduction of LNG in the European scenario. The authors conclude that the technology for transporting natural gas (NG) is mature and expanding through the use of compressed form in urban and light vehicles. The introduction of LNG can expand the natural gas use over long distances due to its higher energy density. In addition, the use of LNG in HDV reduces GHG emissions per kilometre by up to 20% and eliminates nearly 100% of sulphur oxides and particulate matter, and also reduces the noise level in cities compared to diesel trucks. In addition, according to the authors, forecasts for the introduction of LNG in Spanish road haulage could reduce GHG emissions by 12% and diesel fuel consumption by 42% in the long term. The study [15] is devoted to an experimental study of the possibilities of reducing GHG emissions and improving the energy efficiency of engines using NG as the main fuel.

The costs associated operation of buses and municipal vehicles with the gas-engine were analyzed in the study [16]. Noting the advantages of using CNG compared to diesel fuel, the study's author [17] believes that the high vehicle costs and undeveloped infrastructure may hinder their deployment. Review [18] shows that CNG and LNG are important for determining the development of vehicles powered by natural gas. But at the same time, according to the authors, additional measures are needed in the form of subsidizing infrastructure development, encouraging the use of gas-powered vehicles by fleets owners, etc.

2.2 Organizational and Managerial Measures as a Way to Improve the Environmentally Friendly Transport

Organizational and managerial measures can be divided into those that are taken during transportation planning (for example, choosing a rational route, maximizing vehicle load, etc.); the second measures category is related to the human factor affecting driving quality (eco-driving, driving style, safety regulations abidance, etc.). A number of studies have also been carried out in these areas, which show that the use of evidence-based management methods helps to reduce the negative impact on the environment.

Routing vehicles in cities is one of the ways to redistribute traffic in the urban road network, which contributes to more economical and environmentally friendly driving. Search for optimal routes can be done in different ways. Thus, computational experiments conducted by the authors of the article [19] on a number of test cases, using a reference examples set from the literature, show that their proposed hybrid genetic algorithm (HGA) has surpassed modern methods from the literature in almost all cases. The HGA consists of a genetic algorithm combined with local search and capability determination procedures. The article authors [20] developed a mixed integer linear programming (MILP) model for the multiple Capacitated Arc Routing Problem (CARP) task to minimize total costs. Article [21] developed a new mathematical model for reliable Periodic Capacitated Arc Routing Problem (PCARP), which takes into account multiple trips and drivers, as well as the crew working time to study the uncertain nature of the demand parameter. The proposed model's goal function is aimed at minimizing the total distance travelled and the of using vehicles' total cost during the planning period.

A number of studies have been devoted to the study of eco-driving impact on emissions [22, 23]. Thus, in article [24], the influence of eco-driving is considered from two sides: the author takes into account not only the generally accepted goal of reducing fuel consumption, but also reducing emissions of pollutants; i.e. the work compares two different cases: the economic and ecological vehicles operation. In the research course, the author found that the economic as well as ecological behaviour in the eco-driving field shows a significant decrease in energy consumption due to a better speed choice and acceleration rate. The author claims that the algorithm developed and the results will be used in the development of an effective driver assistance system for eco-driving. Similar studies are presented in article [25], which summarizes the factors affecting the eco-driving improvement for drivers, based on an data collected analysis from a social experiment conducted in Toyota City in October 2009 and January 2010. The authors note that, as a tool for promoting eco-driving, some driver assistance systems have been developed to improve driver methods by providing information after evaluating driver behaviour while traveling.

Fuel consumption and fuel cycle emissions were compared by the article authors [26] on controlled driving cycles and their modified eco-driving cycles. Eco-driving savings were compared to reductions in fuel emissions and emissions expected by converting transit fleets to compressed natural gas (CNG), another popular fuel conservation strategy. The results showed that eco-driving would be a potentially very economical strategy for local and express transit.

2.3 The Use of Measures to Improve Sustainability: Case Study of the Garbage Trucks Work

The above described directions for improving environmental friendliness are more applicable to large fleets, for which both organizational-managerial and technical-technological methods can be more targeted and bring tangible effect. One of such examples can be a municipal vehicles fleet: street cleaning vehicles and garbage trucks. All over the world, city's habitant pay attention to the environmental friendliness and noiselessness of communal machines. To reduce harmful emissions, the transfer of municipal vehicles to a more environmentally friendly types of fuel. The noise associated with traffic near residential buildings, especially caused by garbage trucks in the morning, has a negative effect on the health and general emotional state of city's habitant. Gas-engine garbage trucks are less noisy than diesel ones, which makes their work less noticeable for residents.

In addition, since such vehicle moves mainly along permanent routes, scientific routing methods can be applied. In papers [27, 28], optimization of drawing up the garbage trucks route to improve the ecological situation and fuel costs was considered. Route management often requires adjustments due to traffic congestion, sudden equipment failures, repair of road sections, etc. As an additional equipment on modern garbage trucks, video cameras are used to facilitate manoeuvrings the vehicle in reverse when feeding vehicles to the garbage can, automatic's compactor's weighing systems, the GLONASS or GPS navigation system with the ability to exchange information with the central carrier's dispatching station.

To solve the problem of managing garbage disposal, the article authors [29] suggest using garbage cans with IoT support. These garbage cans use RFID tags to track waste associated with a networked web-based system, and according to the weight of the added waste, the host server calculates points and updates in the virtual wallet database. It also measures the fullness of the recycle bins and updates the status of each recycle bin on the municipal server. The article authors [30] believe that since the garbage truck route affects fuel consumption, as well as the working time of an employee, it is necessary to solve this problem for each specific case. The Pigeonhole algorithm is used to select a garbage truck, after which the data generated from the Pigeonhole is then used to determine the shortest path using the Dijkstra algorithm.

3 Results and Discussion

The main consumers of cargo gas-engine vehicles are commercial carriers, food industry's transport and logistics services and trade, oil and gas industry, construction, public utilities. For urban municipal services, gas-engine vehicles are most relevant, since the replacement of diesel analogues by gas-engine ones will improve the environmental situation in large cities. In addition, since such vehicles is operated on well-known and short routes, refuelling problems can be avoided. This automotive market segment was developed thanks to the state program for subsidizing gas-engine fleet.

3.1 Market Structure of Municipal Vehicles in Russia

Today, the main producers of municipal vehicles in Russia are the leading machine-building enterprises, among which are: OJSC "Arzamas Plant of Municipal machine-building" (Arzamas, Nizhny Novgorod Region); OJSC "Ryazhsky auto repair plant" (Ryazhsk, Ryazan region); OJSC "Mtsensk Plant of Municipal machine-building" (Mtsensk, Oryol Region). They account for 78.5% of the garbage truck market. Despite the relatively high production volumes of Municipal vehicles by Russian enterprises, the import of Municipal vehicles on the Russian market plays an important role. According to the results of January–November 2017, 1037 Municipal vehicles with a total value of $ 2 million entered the Russia territory [31]. The Municipal vehicles import basis is Chinese-made vehicles, Germany, are in second place among the manufacturing countries, followed by the Netherlands and Italy.

Modern Russian garbage trucks are mostly produced on KAMAZ and MAZ chassis. Foreign chassis manufacturers also aspire to the garbage truck segment. To reduce the cost, they adapted their chassis for assembling on them superstructures of Russian automobile plants, and also organized the release of the most modern models of their equipment at Russian enterprises. The Mtsensk Plant of Municipal machine-building and the Ryazhsky Car Repair Plant adapted their equipment to the Ford Cargo and HUYNDAI chassis. The experimental repair and mechanical plant "Spetstrans" installs containers on the Isuzu, TATA chassis.

The recent year's trend is the increase in the share of garbage trucks with rear loading. They can be used both for removal of solid household waste from waste transfer stations to landfills, and for working in yards in conditions of dense urban development. The most popular models of rear-loaded garbage trucks are presented in Table 1.

Table 1. Technical characteristics of the most popular garbage truck models with rear loading.

Model	KO-440B	RG 35	KO-427-73	KO-427-34	MKZ-3402
Special equipment plant	Arzamas plant of municipal machine-building	"RG Techno", Moscow region, Lyubertsy	Mtsensk Plant of Municipal machine-building		Ryazhsky vehicles repair plant
Base chassis	KAMAZ-53605	Scania P360	MAZ-5340V2	MAZ-5340V2	MAZ-5337A2
Body capacity, m3	16	24	18,5	16	18 ± 0,2
Weight of waste loaded, kg	7200	12000	7300	7300	5920
Garbage compacting ratio	2,5 ÷ 9	1 ÷ 7	< 6	< 6	K 6
Loading capacity of the manipulator, kg	800	no data	700 (2200)	700 (2200)	500
Loading bucket capacity, m^3	2	2,2	1,1 (8)	1,1 (8)	2,0
Gross weight, kg	20500	31000	19500	19500	18000
Overall dimensions, mm					
- length	8500	9686	8300	8300	8400
- width	2550	2550	2550	2550	2500
- height	3700	3631	3800	3500	3600

3.2 DSS to Control Reliability and Safety of Gas-Cylinder Municipal Vehicles

Every year, the automobile factories' product undergoes significant changes, both the design and software become more complex, which requires careful preparation both in production and in operation. The peculiarity of the municipal vehicles is that the chassis is made by one manufacturer, and the special equipment is made by another. Therefore, separate maintenance regulations are required. Moreover, the presence of gas equipment complicates the task of ensuring the smooth operation of such gas-engine vehicles. There is another problem on the manufacturer part - the reliable statistics collection on faults. It is also important that, based on the analysis and evaluation of this data, be able to refine the vehicle's systems, components or aggregates. This area of activity acquires particular relevance with the increasing complexity of the design; therefore, the single information space's organization between the manufacturer and fleet owners plays an important role in maintaining the such vehicles productivity and preventing failures. The interaction between manufacturers and owners through special control centres will allow to quickly identify problems and solve them [32]. Such opportunities are currently opened by the systems digitalization, which allows to create a single information space between the production system and the systems of operation and service.

For interactive control and diagnosis of systems, aggregates and components of chassis and special equipment when performing their functional tasks, they must be equipped with sensors and actuators. This is necessary to obtain diagnostic information and process it to issue a diagnostic report. The on-board computer integrates the information received and informs the driver about the need to perform certain actions. At certain times, information about the condition of the vehicle and its location goes to the server in the managing centre. The diagnostic algorithm construction of the is preceded by the statistical data analysis to identify the most frequently occurring faults and failures. Based on the prediction results, the server gives permission for further vehicle operation or, for critical processing results, initiates a message to the driver about the need for maintenance or an emergency vehicle stop (Fig. 1).

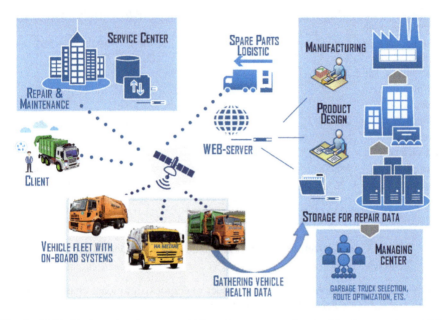

Fig. 1. Unified informational system of life cycle support to the municipal gas-engine vehicle

3.3 Securing Garbage Trucks on CNG

The process of gas-engine vehicles maintaining with in working condition has its own features. Failure-free working of gas-balloon equipment (GBE) during the entire operation life is impossible without complying with the complex of organizational and technical measures aimed primarily at the owner safety and the surrounding space. The main requirements for the elements of vehicles special equipment whose engines run on compressed natural gas are given in the UNECE Regulations [33].

The gas cylinders' production and operation are regulated by special regulatory documents, including GOST R ISO 11439-2010 "Gas cylinders. High pressure cylinders for storage natural gas as a fuel on the vehicle. Technical conditions" [34]. The main component of gas equipment for methane are gas cylinders. They are of four types: metal; metal, covered with fiberglass liner; metal-composite; composite. Then balloon is heavier, the it is less effective, because the tare coefficient is higher. However, then balloon is lighter, the it is more expensive. The question is what is more important for the vehicle's owner: to reduce the gas equipment cost by sacrificing a vehicle's commercial load's part, or to minimize the additional weight, while paying more.

Numerous tests of vehicles operating on CNG have confirmed this technologies' high safety level. Even in serious accidents, when the vehicle received significant damage, the gas equipment remained tight and fully functional. To ignite natural gas, it is necessary that its concentration in the air be from 5 to 15%. In addition, it has a very high auto-ignition temperature: 540 °C. Comparison of natural gas in these indicators with propane, gasoline and diesel fuel shows that it is the safest motor fuel type. It is not uncommon when, after a serious accident or fire, gas cylinders remained unharmed. In such situations, a safety system was triggered (this is a valve with a bursting disc, which is triggered by a sharp increase in pressure, as well as a valve with a fusible insert, which is triggered by an increase in temperature) and the natural gas is released into the atmosphere.

3.4 Ensuring the Reliability of Garbage Trucks on CNG Through the Service System

As noted above, in order to develop the market for gas-engine vehicles on CNG, it is necessary to ensure the development of infrastructure: gas filling stations and fuel delivery systems, diagnostic and service centres, as well as trained personnel, because we have quality problems to gas-engine vehicles' service. This is due to the technological processes specifics of maintenance and current repair (M & R) of gas-engine vehicles. The service centre in its territory should have:

- workstation for leak test on gas equipment;
- workstation for gas release (accumulation) and degassing cylinders;
- a specialized area for gas equipment's M & R;
- warehouse for emptied degassed cylinders' storage for CNG;
- open storage areas for CNG vehicles.

Before being put on the maintenance station, the tightness of the gas equipment and its connections is checked. In case of gas leakage in the joints, entry into the GBE maintenance station is allowed only after the leak has been removed. At the end of the gas equipment maintenance or repair, the vehicle is sent to general stations for further vehicle's M & R as a whole. In the course of maintenance during the main operation period, the works displayed in the algorithm shown in Fig. 2 are performed.

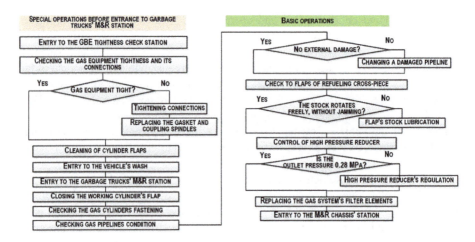

Fig. 2. Technological process of gas equipment maintenance

It is proposed to apply Failure Modes and Effects Analysis (FMEA) for planning measures to improve the quality of maintenance of gas-cylinder equipment for municipal vehicles. FMEA is an analysis method used in quality management to identify potential defects (inconsistencies) and the reasons for their occurrence in a product, process or service [35, 36].

In accordance with this method, any defect (inconsistency) of the considered analysis object can be quite fully characterized by three criteria:

- Rank of significance (S) - scoring the nonconformity consequences severity.
- Occurrence rank (O) - a scoring of the nonconformity's cause frequency (probability).
- Detection rank (D) - scoring the probability of detecting a potential defect or its cause.

For each of these criteria, there is a expert assessments scale in the range from 1 to 10. Moreover, this scale is increasing, that is, the greater the significance or frequency of failure, the higher the corresponding estimates [37].

The integral (generalized) assessment of the certain defect criticality is calculated as the three assessments product according to the three specified criteria and represents the priority risk coefficient (PRC). This generalized assessment can take values from 1 to 1000, and, the higher it is, the more damage this defect (non-conformity) can cause. For example, an insufficiently high quality of gas equipment maintenance may have the following consequences:

- threat to consumer safety;
- reduced customer satisfaction due to financial losses
- reduction of the company's image, loss of trust in the brand.

The quality of maintenance of gas-engine municipal vehicles depends on numerous factors, among which there are cause-effect relationships. In the FMEA process, the problems causes are divided into key categories [38]. These categories are human, technology, equipment, materials (inventory) and the environment. All the reasons related to the problem under study are detailed within these categories frame:

- Causes related to a person include factors due to the person's condition and capabilities.
- Technology-related reasons encompass how the work is done, as well as everything related to the productivity and accuracy of the operations, process or actions.
- The reasons related to the equipment are all factors that are caused by the equipment, devices used in the actions implementation.
- Inventory-related causes are all factors associated with consumables and components.
- The reasons related to the external environment are all factors that determine the impact of the external environment on the actions performance [38].

All factors were analyzed and ranked according to the influence degree on the maintenance process of gas-engine vehicles. The ranking result of the consequences and causes of maintenance insufficient quality is presented in Table 2.

Table 2. FMEA matrix of gas-engine vehicles' maintenance system.

Potential discrepancy	Consequences	S	Causes of potential breach	O	D	PRC
Insufficient GBE maintenance quality	threat to consumer safety	10	Non-compliance with maintenance technology	3	3	90
			Insufficient capacity for a quality technology process	7	3	210
	reduced customer satisfaction due to financial losses	7	Inadequate staff qualifications	2	3	60
			Insufficient staff motivation	3	3	90
			Incomplete staffing	2	1	20
			Insufficient lighting in the workstatios	1	2	20
			Violation of temperature and humidity conditions	2	2	40
			Insufficient cleanliness and order in the workstatios	2	2	40
	reduction of the company's image, loss of trust in the brand	7	Outdated equipment	3	3	90
			Late provision of inventory items;	6	3	180
			Low quality of supplied spare parts;	5	3	150

As it can be seen from Table 2, the following factors have the greatest influence on the GBE maintenance quality:

- Late provision of inventory items;
- Low quality of supplied spare parts;
- Insufficient capacity for a quality technology process.

Since these reasons are the most significant, corrective and preventive actions should be assigned, first of all, to correct these factors (Fig. 3).

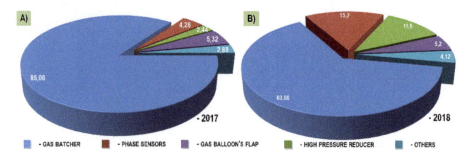

Fig. 3. Shares of gas equipment's reclamations ((A) - before to the measures implementation to improve processes; (C) after the measures implementation)

Table 3. FMEA matrix of the gas equipment maintenance's technological process.

№	Process	Possible defect	Consequences	S	Detection measures	D	PRC
1	Checking the tightness of gas equipment and its connections	Breach impermeability of gas equipment and its connections	Fire and explosion hazard	10	Leak detector	1	70
2	Checking the fastening of gas cylinders on the bracket	Loose cylinder mounting	Axial displacement and turning cylinders. Loss of cylinders. Tube rupture	9	Visual inspection	1	90
3	Checking the condition of gas pipelines	Damage to gas pipelines	Leak of gas	9	Visual inspection	1	90
4	Check to flaps of refueling cross-piece	Rod rotates with jamming	Gas leakage when filling cylinders	9	Visual inspection	1	90
5	Check high pressure reducer	Sharp decrease in pressure at the exit from a reducer	Insufficient fuel supply. Difficult engine start	7	Tescan programmer Askan-10	2	98

(*continued*)

Table 3. (*continued*)

№	Process	Possible defect	Consequences	S	Detection measures	D	PRC
6	Gas meter check	Jamming dispenser; Dispenser's Electromagnet Failure	Significant deviation of the composition of the air-fuel mixture. Insufficient fuel supply. Difficult engine start	7	Not checked	7	343
7	Phase sensor check	No phase sensor signal	Minor deterioration of power indices. Slight increase in fuel consumption and exhaust emissions	6	Not checked	7	294
8	Cylinder valve check	Damage to the valve body; Worn O-rings	Leak of gas	9	Not checked	7	441

To identify opportunities for technological process improving, potential defects types in gas-balloon equipment that arise when vehicle's operating in the millage range of 25–30 thousand km, as well as their implications for the consumer safety and comfort, were studied. The data was obtained in the production and technical service of one of the enterprises servicing and operating municipal vehicles on gas engine fuel. The FMEA results are presented in Table 3.

The FMEA results served as the basis for correction the technological process of gas equipment M & R. This has reduced the failures number in the gas inflow system, by improving the processes quality and preventing sudden failures, which increases the reliability and safety of gas-engine municipal vehicles' operation. The improved process is shown in Fig. 4.

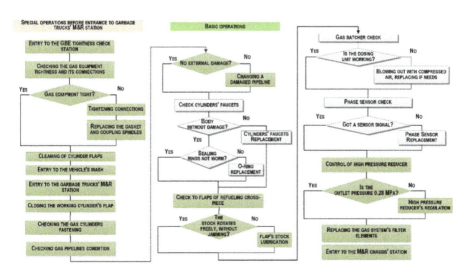

Fig. 4. Improved technological process of maintenance of gas equipment

4 Conclusion

The difficult environmental situation caused by the harmful effects of emissions from diesel and gasoline in cities has made CNG a very promising alternative fuel for municipal transport. The use of natural gas as a transport fuel for municipal vehicles can help improve the environmental situation in cities, reduce the harmful effects on health by reducing noise pollution and reducing harmful emissions in the atmospheric air.

The safety of using vehicles on CNG around the world is well established and commercially available for all types of vehicles. The main directions of global garbage truck market development are the development of garbage trucks on a special chassis with a low-entry cab, front-loading garbage trucks (front-loader), garbage trucks based on a chasses with gas or hybrid engine, garbage trucks with the possibility of selective garbage collection. In Europe, the production and use of garbage trucks operating on alternative fuels is growing every year. Garbage collection companies are increasingly choosing cars with hybrid drives or engines running on electricity, compressed natural gas or biomethane. Problems for routes optimizations garbage trucks and eco-driving Solved by many scientists and engineers. For these purposes, they use intelligent systems and methods, as well as the principles of the Internet of things.

However, the important issue remains the trouble-free operation of municipal vehicles. The article shows that the creation of a unified information management system will facilitate the timely receipt of adequate information about failures and the improvement of the M & R process based on its data analysis. The article used the FMEA method for this purpose. Using the example, the authors show that the improvement of the service process makes it possible to improve their quality, as well as the reliability and vehicles reliability, hence the sustainability and safety of the entire transport system.

References

1. Air Pollution in Russia: Real-time Air Quality Index Visual Map. http://aqicn.org/map/russia/
2. Khan, M.I., Yasmeen, T., Farooq, M., Wakeel, M.: Research progress in the development of natural gas as fuel for road vehicles: a bibliographic review (1991–2016). Renew. Sustain. Energy Rev. **66**, 702–741 (2016)
3. Yang, W., Yu, C., Yuan, W., Zhang, X.W., Wang, X.: High-resolution vehicle emission inventory and emission control policy scenario analysis, a case in the Beijing-Tianjin-Hebei (BTH) region China. J. Clean. Prod. **203**, 530–539 (2018)
4. Liu, Y.-H., Liao, W.-Y., Lin, X.-F., Li, L., Zeng, X.: Assessment of Co-benefits of vehicle emission reduction measures for 2015–2020 in the Pearl river delta region, China. Environ. Pollut. **223**, 62–72 (2017)
5. Song, C., Ma, Ch., Zhang, Y., Wang, T., Wu, L., Wang, P., Liu, Y., Li, Q., Zhang, J., Dai, Q., Zou, Ch., Sun, L., Mao, H.: Heavy-duty diesel vehicles dominate vehicle emissions in a tunnel study in northern China. Sci. Total Environ. **637–638**, 431–442 (2018)

6. Huang, Ch., Tao, Sh, Lou, Sh, Hu, Q., Wang, H., Wang, Q., Li, L., Wang, H., Liu, J., Quan, Y., Zhou, L.: Evaluation of emission factors for light-duty gasoline vehicles based on chassis dynamometer and tunnel studies in Shanghai, China. Atmos. Environ. **169**, 193–203 (2017)
7. Song, H., Ou, X., Yuan, J., Yu, M., Wang, C.: Energy consumption and greenhouse gas emissions of diesel/LNG heavy-duty vehicle fleets in China based on a bottom-up model analysis. Energy **140**(1), 966–978 (2017)
8. Arteconi, A., Brandoni, C., Evangelista, D., Polonara, F.: Life-cycle greenhouse gas analysis of LNG as a heavy vehicle fuel in Europe. Appl. Energy **87**, 2005–2013 (2010)
9. ISO 14040. Environmental management – life cycle assessment – principles and framework. International Organisation for Standardisation, Geneva (2006)
10. ISO 14044. Environmental management – life cycle assessment – requirements and guidelines. International Organisation for Standardisation, Geneva (2006)
11. Makarova, I., Shubenkova, K., Mavrin, V., Gabsalikhova, L., Sadygova, G., Bakibayev, T.: Problems, risks and prospects of ecological safety's increase while transition to green transport. In: Nathanail, E., Karakikes, I. (eds.) CSUM 2018. Advances in Intelligent Systems and Computing, vol. 879, pp. 172–180. Springer, Cham (2019)
12. Sukumaran, R.K., Singhania, R.R., Mathew, G.M., Pandey, A.: Cellulase production using biomass feed stock and its application in lignocellulose saccharification for bio-ethanol production. Renew. Energy **34**, 421–424 (2009)
13. Altın, R., Çetinkay, S., Yücesu, H.S.: The potential of using vegetable oil fuels as fuel for diesel engines'. Energy Convers. Manage. **42**(5), 529–538 (2001)
14. Osorio-Tejada, J.L., Llera-Sastresa, E., Scarpellin, S.: Liquefied natural gas: Could it be a reliable option for road freight transport in the EU? Renew. Sustain. Energy Rev. **71**, 785–795 (2017)
15. Kozlov, A.V., Terenchenko, A.S., Luksho, V.A., Karpukhin, K.E.: Prospects for energy efficiency improvement and reduction of emissions and life cycle costs for natural gas vehicles. In: IOP Conference Series: Earth Environment Science 52, conference 1 (2017)
16. Johnson, C.: Business case for Compressed Natural Gas in municipal fleets. National, Renewable Energy Laboratory. Technical Report NREL/TP-7A2-47919 (2010)
17. Kragha, O.C.: Economic implications of natural gas vehicle technology in U.S. private automobile transportation. https://dspace.mit.edu/handle/1721.1/59686
18. Hao, H., Liu, Z., Zhao, F., Li, W.: Natural gas as vehicle fuel in China: a review. Renew. Sustain. Energy Rev. **62**, 521–533 (2016)
19. Arakaki, R.K., Usberti, F.L.: Hybrid genetic algorithm for the open capacitated arc routing problem. Comput. Oper. Res. **90**, 221–231 (2018)
20. Tirkolaee, E.B., Alinaghian, M., Hosseinabadi, A.A.R., Sasi, M.B., Sangaiah, A.K.: An improved ant colony optimization for the multi-trip capacitated arc routing problem. Computers and Electrical Engineering, in press (2018)
21. Tirkolaee, E.B., Mahdavi, I., Esfahani, M.M.S.: A robust periodic capacitated arc routing problem for urban waste collection considering drivers and crew's working time. Waste Manage. **76**, 138–146 (2018)
22. Andrieu, C., Pierre, G.S.: Comparing effects of eco-driving training and simple advices on driving behavior. Procedia - Soc. Behav. Sci. **54**, 211–220 (2012)
23. Ho, S.-H., Wong, Y.-D., Chang, VW.-Ch.: What can eco-driving do for sustainable road transport? Perspectives from a city (Singapore) eco-driving programme. Sustainable Cities and Society **14**, 82–88 (2015)
24. Mensing, F., Bideaux, E., Trigui, R., Ribet, J., Jeanneret, B.: Eco-driving: an economic or ecologic driving style? Transp. Res. Part C **38**, 110–121 (2014)
25. Ando, R., Nishihori, Y.: A study on factors affecting the effective eco-driving. Procedia – Soc. Behav. Sci. **54**, 27–36 (2012)

26. Xu, Y., Li, H., Liu, H., Rodgers, M.O., Guensler, R.L.: Eco-driving for transit: an effective strategy to conserve fuel and emissions. Appl. Energy **194**, 784–797 (2017)
27. Karimipour, H., Tam, V.W.Y., Burnie, H., Le, K.N.: Vehicle routing optimization for improving fleet fuel efficiency: a case study in Sydney, Australia. Int. J. Environ. Sci. Dev. **8**(11), 776–780 (2017)
28. Król, A., Nowakowski, P., Mrówczynska, B.: How to improve WEEE management? novel approach in mobile collection with application of artificial intelligence. Waste Manage. **50**, 222–233 (2016)
29. Mirchandani, S., Wadhwa, S., Wadhwa, P., Joseph, R.: IoT enabled dustbins. In: 2017 International Conference on Big Data, IoT and Data Science (BID), pp. 73–76. IEEE Press, New York (2017)
30. Hartatik, H., Purbayu, A., Triyono, L.: Dijkstra methode for optimalize recommendation system of garbage transportation time in surakarta city. IOP Conf. Ser.: Mat. Sci. Eng. **333**(1), 012106 (2018)
31. Russian market of utility vehicles. https://os1.ru/article/17601-rossiyskiy-rynok-kommunalnoy-tehniki [in Russian]
32. Makarova, I., Khabibullin, R., Belyaev, E., Belyaev, A.: Improving the system of warranty service of trucks in foreign markets. Transp. Probl. **10**(1), 63–78 (2015)
33. UNECE LNG Vehicles Regulation Adopted. https://www.lngworldnews.com/unece-lng-vehicles-regulation-adopted/
34. GOST R ISO 11439-2010 Gas cylinders. High pressure cylinders for storage on the vehicle of natural gas as a fuel. Technical conditions. http://docs.cntd.ru/document/1200085522 (in Russian)
35. How Failure Mode and Effects Analysis (FMEA) is Used in the Auto Industry. https://www.brighthubpm.com/monitoring-projects/47746-fmea-in-the-automotive-industry/
36. A New Approach to FMEA in the Automotive Industry. http://www.theauditoronline.com/a-new-approach-to-fmea-in-the-automotive-industry/
37. Poprocký, R., Stuchlý, V., Galliková, J., Volna, P.: FMEA analysis of combustion engine and assignment occurrence index for risk valuation. Diagnostyka **18**(3), 99–105 (2017)
38. Bellstedt, S.: Fixing a broken maintenance strategy: PM optimization and FMEA. https://www.fiixsoftware.com/blog/fixing-a-broken-maintenance-strategy-pm-optimization-and-fmea/

Electric Vehicles - Problems and Issues

Elżbieta Macioszek(✉)

Transport Systems and Traffic Engineering Department, Faculty of Transport,
Silesian University of Technology, Katowice, Poland
`elzbieta.macioszek@polsl.pl`

Abstract. Electric vehicles have been gradually growing in popularity. There are countries, such as China, USA and some European countries in particular (including the United Kingdom, Norway, Germany, the Netherlands, Sweden, France), where the number of electric vehicles has been increasing on a year to year basis. In Poland, on the other hand, the market of electric vehicles is still very immature, the proof of which can be perceived in the lack of vehicle charging infrastructure or the negligible demand for such services in places where this infrastructure has actually been built, not to mention the minute sales of electric vehicles. This paper addresses results of an analysis of problems and issues connected with electric vehicles, both worldwide and in Poland.

Keywords: Electromobility · Electric vehicles · Battery-electric vehicles · Plug-in hybrid electric vehicles · Charging points

1 Introduction

Electric vehicles have been gradually growing in popularity over the recent years. Currently, ca. 500,000 electric cars are annually sold worldwide. According to a Bloomberg New Energy Finance report [1], in 2040, 54% of all newly sold cars will be electric cars, which will represent 33% of all vehicles using roads around the world. The Polish draft law on electromobility and alternative fuels [2] defines an electric vehicle as an internal combustion electric vehicle featuring built-in batteries or using only electric energy for propulsion purposes, where electric energy is accumulated by connecting to an external power source.

The development of electromobility is inextricably linked with technological development in general. One of the main obstacles preventing the widespread use of electric cars is currently the properties of batteries. Prices of batteries vary depending on the parameters of their storage systems and operating models. Consequently, the prices of batteries for passenger cars differ significantly from those intended for buses. It is assumed that in 2020 the battery technology will be sufficiently advanced to overcome one of the main drawbacks of the contemporary electric cars, namely their limited operating range. Moreover, batteries are likely to be much cheaper than at present. A decrease in battery prices will be one of the factors expected to reduce the difference between the purchase price of an electric vehicle and that of an internal combustion vehicle. According to the Bloomberg New Energy Finance report [1], 50%

of an electric vehicle price currently accounts for the cost of battery production. In 2020, as technology advances, this cost is assumed to fall to ca. 36%.

Thanks to ongoing and multifaceted scientific research in the field of development [3–7] and upgrading [8–13] of transport systems and networks, one can observe their dynamic development. Some important aspects determining the development of transport systems include organisation, management and information [14–25]. In terms of mobility improvement, on the other hand, electric vehicles are perceive as the next stage of development. This paper addresses the problems and issues connected with electric vehicles, both worldwide and in Poland.

2 Electric Vehicles in Poland

According to the data published by the European Alternative Fuels Observatory [26], there were 290 electric vehicles (passenger cars) registered in Poland in 2017, including 239 battery electric vehicles and 51 plug-in hybrid electric vehicles. The battery electric category is dominated by such vehicles as Hyundai Ioniq Electric, Nissan Leaf, BMW i3, Tesla Model S, Tesla Model X, Renault Zoe and Nissan e-NV200 Evalia (Fig. 1). The most popular plug-in hybrid electric vehicles are BMW i3 Rex, Mitsubishi Outlander PHEV and BMW i8 (Fig. 2).

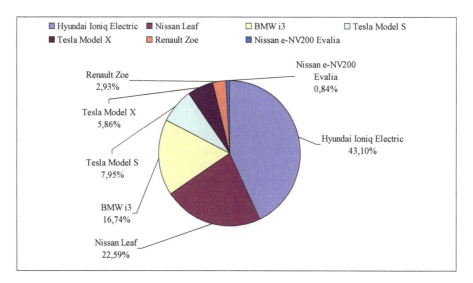

Fig. 1. Share of individual car brands in the group of battery electric vehicles (source: own on the basis of [26])

Electric vehicles are still a rarity rather than a standard element of the car fleet operated in Polish towns. The market of electric vehicles is still germinating in Poland, the proof of which can be perceived in the lack of vehicle charging infrastructure or the negligible demand for such services in places where this infrastructure has actually been built, not to mention the minute sales of electric vehicles. A transition to the next phase of market development will require specific conditions to be created to enable change in many areas. According to a report by the Polish Automotive Industry Association concerning the automotive industry [27], there were 10,400 alternative fuel cars registered in Poland in 2016. It was nearly 76% more than in 2015. The vast majority are hybrid vehicles (95%), However, these vehicles still account for a negligible share in Poland's total automotive market: 2.4% attributable to hybrid vehicles, and 0.1% for electric or plug-in hybrid vehicles (with optional socket charging or station charging). Therefore the number of electric cars used in Poland is completely incomparable to the number of electric cars operated in other countries, for instance in Norway, where a policy dedicated to encouraging residents to buy electric cars has been implemented for many years. According to [28], ca. 20% of all vehicles operated in Norway in 2017 (i.e. slightly more than 1,000 vehicles) were electric vehicles. However, one should also note that at the same time electric cars accounted for 42% of all newly registered cars in Norway in 2017.

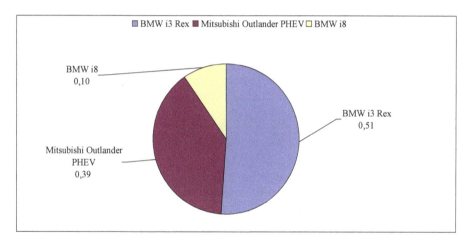

Fig. 2. Share of individual car brands in the group of plug-in hybrid electric vehicles (source: own on the basis of [26])

There are two official documents currently functioning in Poland, namely the Electromobility Development Plan [29] and the National Policy Framework for the Development of Alternative Fuels Infrastructure [30], that provide grounds for the continuous work on the act on electromobility and alternative fuels conducted at the Ministry of Energy.

According to the said Plan, the electromobility development should proceed in three phases, depending on the degree of the market maturity and the required involvement of the state, which are as follows:

- Phase one of preparatory nature, scheduled to proceed until the year 2018. This phase was planned to cover building conditions for the electromobility development in terms of the regulatory framework and establishing rules of public financing.

- Phase two, scheduled for the years 2019–2020, intended to cover building of the infrastructure needed for charging of electric vehicles in selected urban areas and intensifying the incentives to buy electric vehicles. The expectations are that results of the research activities performed in the area of electromobility, commenced already in the first phase, will be commercialised, and new business models aimed to popularise electric vehicles will be implemented.
- Phase three, spanning the years 2020–2025, assuming that the electromobility market will have matured, allowing for the public aid instruments to be gradually withdrawn.

According to the data published by the European Alternative Fuels Observatory [26], there were 329 electric vehicle charging points operating in Poland in 2017. As many as 290 of them were the normal power type stations with the capacity of ≤ 22 kW. Other charging stations (i.e. 39 charging points) enable quick charging of electric vehicles with the capacity exceeding 22 kW. Locations of some of these service points is presented, among others, on an interactive map of electric car charging points available in Poland, prepared by Tesla [31]. The map is available free of charge on the web.

3 Electric Vehicles Worldwide

Electric vehicles were among the first to have been used to transport people in the first half of the 19th century. It was Robert Anderson who had designed an electric-powered carriage prototype. At that time, similar designs of vehicles aimed to replace train horses with electric motors were developed in the USA, the Netherlands, the United Kingdom, France and Hungary, among other countries. Not all of them reached the commercial implementation phase. However, the then manufacturers of electric vehicles were predominantly oriented towards wealthy city dwellers. They were also offering vehicles suitable for women in terms of the ease of use, visual appeal, cleanliness and low noise levels. At the beginning of the twentieth century, electric vehicles were driven out of the market by vehicles with internal combustion engines. The main reason for this phenomenon was the technological limitations of the contemporary power supply technology, enabling these vehicles to develop only relatively low speeds. Electric vehicles also fell into disfavour due to the mass production of Ford T between 1908 and 1912. Moreover, the continuous improvement of internal combustion vehicles resulted made them significantly more practical than electric vehicles. Not until the 1980's had electric vehicles regained popularity, when a policy promoting this sphere was implemented in the USA (electric vehicles were officially recognised as zero-emission means of transport). A synthetic history of the continuous effort related to electric vehicles, illustrated by numerous examples, was created by P. Barycki [32].

The energy used in an electric vehicle to drive an electric motor may be:

- stored in batteries,
- stored in batteries charged by generators,
- produced by an internal combustion engine,

- produced directly by the reactions taking place in fuel cells,
- obtained from a direct connection to the power grid (e.g. trolley buses, trams, railways).

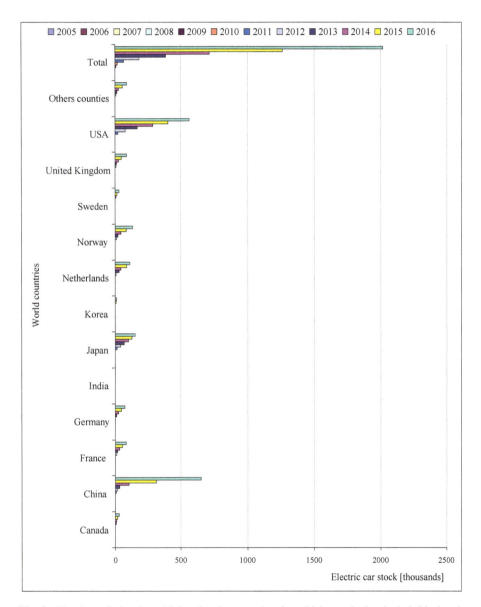

Fig. 3. Number of electric vehicles (i.e. battery electric vehicles and plug-in hybrid electric vehicles) in individual countries in the years 2005–2016 (source: own materials based on [33, p. 49])

A major obstacle preventing the spread of electric vehicles is their low affordability. One account of the limited use, the technologies applied in electric vehicles are much more expensive than combustion technologies. However, one can observe a clear downward trend in terms of the prices of individual components. The factor that will probably cause a significant drop in the prices of electric vehicles in upcoming years will be their growing popularity, as the cost of designing and implementing further solutions will be spread over an increasing number of customers.

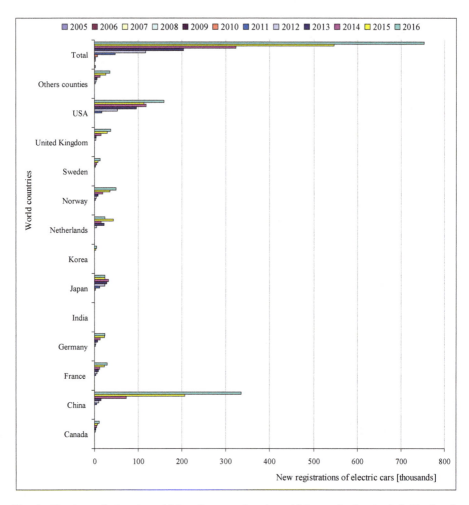

Fig. 4. Number of electric vehicles (battery electric vehicles and plug-in hybrid electric vehicles) newly registered in individual countries in the years 2005–2016 (source: own materials based on [33, p. 50])

The number of electric vehicles (i.e. battery electric vehicles and plug-in hybrid electric vehicles) in individual countries in the years 2005–2016 has been shown in

Fig. 3, where the "other countries" category represents such states as: Belgium, Austria, Croatia, Bulgaria, Southern part of Cyprus, Denmark, the Czech Republic, Estonia, Lithuania, Greece, Finland, Iceland, Poland, Ireland, Hungary, Romania, Lichtenstein, Switzerland, Turkey, Latvia, Portugal, Italy, Spain, Luxemburg, Slovakia, Malta and Slovenia. With reference to Fig. 3, one can conclude that the countries with the largest number of electric vehicles are China, USA and Japan. In Europe, on the other hand, the most electric vehicles are used in the Netherlands, Norway, the United Kingdom, France and Germany.

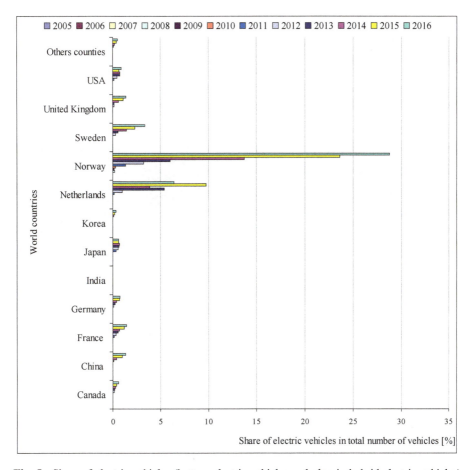

Fig. 5. Share of electric vehicles (battery electric vehicles and plug-in hybrid electric vehicles) in the total number of vehicles in individual countries in the years 2005–2016 (source: own materials based on [33, p. 51])

The statistics of electric vehicles (i.e. battery electric vehicles and plug-in hybrid electric vehicles) newly registered in individual countries in the years 2005–2016 have been illustrated in Figs. 3 and 4. Figure 4 implies that the number of newly registered electric vehicles has been growing year by year. In 2016, the total number of newly registered electric vehicles worldwide in the countries subject to the analysis was below 800,000, of which ca. 336,000 were registered in China and fewer than 16,000 in the USA. In Europe, the most electric vehicles were registered in countries such as the Netherlands, Norway, the United Kingdom, France, Germany and Sweden. It is by no accident that battery electric vehicles and plug-in hybrid electric vehicles are so popular in these countries. Firstly, advanced scientific research on this type of vehicles has been conducted for a long time in all of the above countries, and they are also known of the widespread use of renewable energy and/or nuclear energy. One should note that from the environmental protection perspective, electric vehicles do not emit pollutants in operation. However, on account of the technology applied and the intense use of rare earth elements (rare earth elements being a family of seventeen chemical elements which consists of lanthanides and the scandium series, i.e. scandium, lanthanum, yttrium, praseodymium, cerium, neodymium, samarium, promethium, europium, gadolinium, holmium, terbium, dysprosium, thulium, erbium, lutetium and ytterbium), the process of production of electric cars is a source of considerably more pollutants

Fig. 6. Example of publicly accessible electric vehicles charging station in London (United Kingdom)

Fig. 7. Example of publicly accessible electric vehicles charging station in Budapest (Hungary)

Electric Vehicles - Problems and Issues 177

Fig. 8. Example of electric car rental service in Budapest (Hungary)

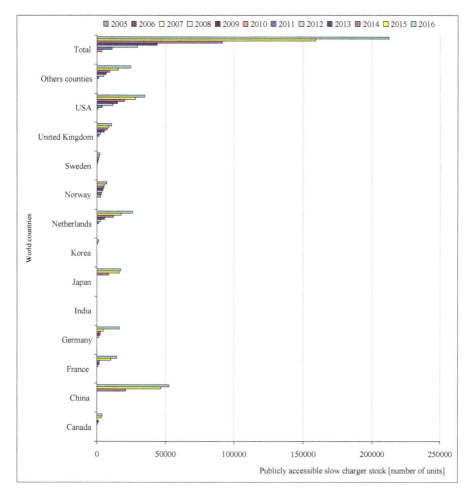

Fig. 9. Stock of publicly accessible slow chargers for electric vehicles in individual countries in the years 2005–2016 (source: own materials based on [33, p. 52])

than the manufacture of internal combustion cars. When ready to be released to the market, an electric car is responsible for emission of ca. 14 tonnes of carbon dioxide (CO_2), while an internal combustion car accounts for ca. 6 tonnes. In countries where the majority of electricity comes from coal, an electric car will be responsible for emission of a comparable amount of carbon dioxide to an internal combustion car over the entire service life. In countries where most of the energy produced is either nuclear and/or renewable, an electric car will emit 24% less carbon dioxide. This is why the countries with the predominant share of nuclear, solar, hydro, geothermal and wind energy are those that report the highest interest in electric vehicles, and consequently the highest rates of their production and registration.

The share of electric vehicles (i.e. battery electric vehicles and plug-in hybrid electric vehicles) in the total number of vehicles in individual countries in the years 2005–2016 has been presented in Fig. 5. Figure 5 implies that this share has been growing year by year. The highest share of electric vehicles in the total number of vehicles is observed in Norway, where it came to nearly 29% in 2016.

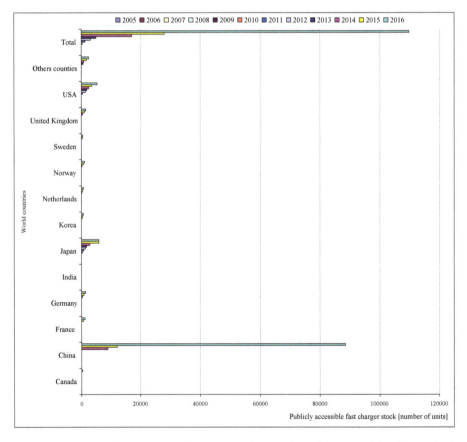

Fig. 10. Stock of publicly accessible fast chargers for electric vehicles in individual countries in the years 2005–2016 (source: own materials based on [33, p. 52])

There are four categories of charging infrastructure [34]:

- public charging station on public domain (e.g. roadside/sidewalk),
- publicly accessible charging station on private domain (e.g. commercial areas such as shopping malls),
- semi-public charging station on public or private domain (e.g. car sharing centre stations, hotels or business parking for visitors and customers),
- privately accessible charging station (e.g. home or office locations).

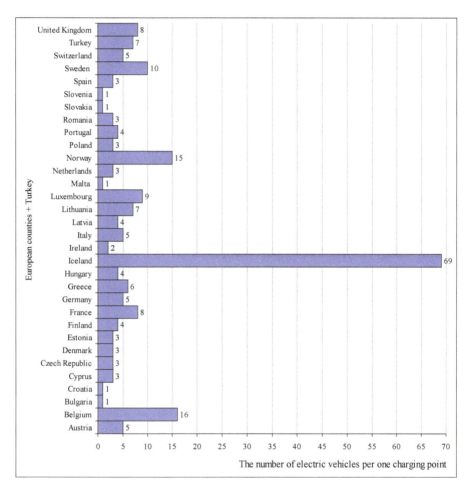

Fig. 11. Number of electric vehicles per a single charging point in European countries and in Turkey in 2017 (total value, i.e. a sum of normal power (≤ 22 kW) and high power (>22 kW) points).

Examples of publicly accessible electric vehicle charging stations and electric car rental services have been presented in Figs. 6, 7 and 8.

According to the data presented in [34, 35], there are nearly 320,000 stations of the publicly accessible charging infrastructure in the world nowadays. The highest number of fast chargers are located in China, and they account for one third of all chargers operating globally. Fast chargers use high-power alternating current, direct current or induction, and can fully recharge a battery of electric vehicles in less than one hour. In 2016, the global average deployment of fast chargers was slower than the overall deployment of chargers (except for China). The stock of publicly accessible slow and fast chargers for electric vehicles in individual countries in the years 2005–2016 has been presented in Figs. 9 and 10. Based on the data provided in Figs. 9 and 10, one can notice that the most charging stations have been installed over the recent years (since 2014) in countries such as China and the USA, while in Europe in the Netherlands, the United Kingdom, France, Germany and Norway.

Figure 11 shows the number of electric vehicles per a single charging point in European countries and in Turkey in 2017 (total value, i.e. a sum of normal power (≤ 22 kW) and high power (>22 kW) points). One will clearly notice that the highest share of vehicles per one charging point is in Iceland (69), while this ratio is the lowest in countries such as Slovenia, Slovakia, Malta, Croatia and Bulgaria (1).

4 Conclusions

In some countries, like China, the USA and some European countries in particular (such as the United Kingdom, Norway, Germany, the Netherlands, Sweden, France), the share of electric vehicles in the total vehicle stock is already significant, and has been systematically growing from year to year. Poland still struggles against a number of major barriers hampering the growth in sales of electric vehicles. The most serious of them is the high cost of purchase of a new electric car (especially when compared to average remuneration), which considerably exceeds that of an internal combustion engine vehicle of a similar class. Consequently, vehicles of this type are still perceived as gadgets rather than fully-fledged utility vehicles. Furthermore, they can cover only a very limited operating range which effectively eliminates the most significant benefit resulting from owning a car, namely the ability to move around the road network without constraints. Additionally, many towns are lacking publically available electric vehicle charging stations or their number of negligible, which significantly reduces the operating range of electric vehicles. Another challenge facing owners of electric vehicles is the long charging time of batteries, which on top of that must be replaced every ten years. On account of all the foregoing characteristics, one may claim it to be reasonable, if not necessary, to undertake further continuous and multifaceted research on electric vehicles.

Acknowledgements. The present research has been financed from the means of the National Centre for Research and Development as a part of the international project within the scope of ERA-NET CoFund Electric Mobility Europe Programme "Electric travelling - platform to support the implementation of electromobility in Smart Cities based on ICT applications".

References

1. Bloomberg New Energy Finance: BNEF's Annual Long-Term Forecast of Global Electric Vehicle Adoption to 2040. https://about.bnef.com/electric-vehicle-outlook/#toc-download
2. Draft Act of 26 of April 2017 on Electromobility and Alternative Fuels. https://legislacja.rcl.gov.pl/docs//2/12297850/12430702/12430703/dokument286622.pdf
3. Małecki, K.: The roundabout micro-simulator based on the cellular automata model. In: Sierpiński, G. (ed.) Advanced Solutions of Transport Systems for Growing Mobility. AISC, vol. 631, pp. 40–49. Springer, Cham (2018)
4. Małecki, K., Wątróbski, J.: Cellular automaton to study the impact of changes in Traffic rules in a roundabout: a preliminary approach. Appl. Sci. Basel **7**(7), Article Number: UNSP 742 (2017)
5. Małecki, K.: The use of heterogeneous cellular automata to study the capacity of the roundabout. In: Rutkowski, L., Korytkowski, M., Scherer, R., Tadeusiewicz, R., Zadech, L. A., Zurada, J.M. (eds.) Artificial Intelligence and Soft Computing. LNAI, vol. 10246, pp. 308–317. Springer, Cham (2017)
6. Macioszek, E.: The comparison of models for follow-up headway at roundabouts. In: Macioszek, E., Sierpiński, G. (eds.) Recent Advances in Traffic Engineering for Transport Networks and Systems. LNNS, vol. 21, pp. 16–26. Springer, Cham (2018)
7. Turoń, K., Golba, D., Czech, P.: The analysis of progress csr good practices areas in logistic companies based on reports "Responsible Business in Poland. Good Practices" in 2010–2014. Sci. J. SUT **89**, 163–171 (2015)
8. Turoń, K., Czech, P., Juzek, M.: The concept of walkable city as an alternative form of urban mobility. Sci. J. SUT **95**, 223–230 (2017)
9. Skrzypczak, I., Kogut, J., Kokoszka, W., Zientek, D.: Monitoring of landslide areas with the use of contemporary methods of measuring and mapping. Civ. Environ. Rep. **24**(1), 69–82 (2017)
10. Skrzypczak, I., Buda-Ozog, L., Pytlowany, T.: Fuzzy method of conformity control for comprehensive strength of concrete on the basis of computational numerical analysis. Meccanica **51**(2, Special Issue SI), 383–389 (2016)
11. Len, P., Oleniacz, G., Skrzypczak, I., Mika, M.: The Hellwig's and zero unitarisation methods in creating a ranking of the urgency of land consolidation and land exchange work. In: Informatics, Geoinformatics and Remote Sensing Conference Proceedings, SGEM vol. II. Book Series: International Multidisciplinary Scientific GeoConference - SGEM, pp. 617–624 (2016)
12. Macioszek, E., Czerniakowski, M.: Road Traffic Safety-Related Changes Introduced on T. Kościuszki and Królowej Jadwigi Streets in Dąbrowa Górnicza Between 2006 and 2015. Sci. J. SUT **96**, 95–104 (2017)

13. Macioszek, E.: The application of HCM 2010 in the determination of capacity of traffic lanes at turbo roundabout entries. Transp. Probl. **11**(3), 77–89 (2016)
14. Golba, D., Turoń, K., Czech, P.: Diversity as an opportunity and challenge of modern organizations in TSL area. Sci. J. SUT **90**, 63–69 (2016)
15. Celiński, I., Sierpiński, G.: Method of assessing vehicle motion trajectory at one-lane roundabouts using visual techniques. In: Macioszek, E., Akçelik, R., Sierpiński, G. (eds.) Roundabouts as Safe and Modern Solutions in Transport Networks and Systems. LNNS, vol. 52, pp. 24–39. Springer, Cham (2019)
16. Celiński, I.: Evaluation method of impact between road traffic in a traffic control area and in its surroundings. Ph.D. dissertation, Warsaw University of Technology Publishing House, Warsaw (2018)
17. Celiński, I.: GT Planner used as a tool for sustainable development of transport infrastructure. In: Suchanek, M. (ed.) New Research Trends in Transport Sustainability and Innovation. SPBE, pp. 15–27. Springer, Cham (2018)
18. Sierpiński, G., Staniek, M., Celiński, I.: Travel behavior profiling using a trip planner. Transp. Res. Procedia **14C**, 1743–1752 (2016)
19. Sierpiński, G.: Distance and frequency of travels made with selected means of transport - a case study for the Upper Silesian conurbation (Poland). In: Sierpiński, G. (ed.) Intelligent Transport Systems and Travel Behaviour. AISC, vol. 505, pp. 75–85. Springer, Cham (2017)
20. Sierpiński, G., Staniek, M.: Education by access to visual information - methodology of moulding behaviour based on international research project experiences. In: Gómez Chova, L., López Martínez, A., Candel Torres I. (eds.) ICERI 2016 Proceedings: 9th International Conference of Education, Research and Innovation, pp. 6724–6729. IATED Academy, Valencia (2016)
21. Sierpiński, G., Staniek, M., Celiński, I.: Shaping environmental friendly behaviour in transport of goods - new tool and education. In: Gómez Chova, L., López Martínez, A., Candel Torres I. (eds.) ICERI 2015 Proceedings: 8th International Conference of Education, Research and Innovation, pp. 118–123. IATED Academy, Valencia (2015)
22. Staniek, M., Czech, P.: Self-correcting neural network in road pavement diagnostics. Autom. Constr. **96**, 75–87 (2018)
23. Chmielewski, J.: Impact of transport zone number in simulation models on cost-benefit analysis results in transport investments. IOP Conf. Ser. Mater. Sci. Eng. **245**, 1–10 (2017)
24. Chmielewski, J.: Transport demand model management system. IOP Conf. Ser. Mater. Sci. Eng. **471**, 1–10 (2019)
25. Chmielewski, J., Olenkowicz-Trempała, P.: Analysis of selected types of transport behaviour of urban and rural population in the light of surveys. In: Macioszek, E., Sierpiński, G. (eds.) Recent Advances in Traffic Engineering for Transport Networks and Systems. LNNS, vol. 21, pp. 27–36. Springer, Cham (2018)
26. European Alternative Fuels Observatory. http://www.eafo.eu/vehicle-statistics/m1
27. Municipal Portal. https://portalkomunalny.pl/po-polsce-jezdzi-coraz-wiecej-aut-elektrycznych-w-2017-r-zarejestrowano-160-elektrykow-356950/
28. Lesman, U.: Od Lat Norwegia Subwencjonuje Przyjazne dla Środowiska Samochody. Rząd w Oslo Uważa, że to Świetna Inwestycja i Nie Zamierza z Nich Rezygnować. http://www.rp.pl/Motoryzacja/161129175-Norwegia-od-lat-intensywnie-promuje-samochody-elektryczne.html
29. Ministerstwo Energii Rzeczypospolitej Polskiej: Plan Rozwoju Elektromobilności w Polsce "Energia do Przyszłości". http://www.me.gov.pl/files/upload/27052/Plan%20Rozwoju%20Elektromobilno%C5%9Bc%20RM.pdf
30. Ministerstwo Energii Rzeczypospolitej Polskiej: Krajowe Ramy Polityki Rozwoju Infrastruktury Paliw Alternatywnych. Warszawa, 29 March 2017. http://bip.me.gov.pl/files/upload/26071/Krajowe_ramy%20_29032017.pdf

31. Interactive Map of Charging Electric Cars in Poland. http://www.mytesla.com.pl/punkty-ladowania/
32. Spiderweb. https://www.spidersweb.pl/e/samochody-elektryczne-historia-powtorka
33. International Energy Agency. https://www.iea.org/publications/freepublications/publication/GlobalEVOutlook2017.pdf
34. Eurelectric - Electricity for Europe: Charging Infrastructure for Electric Vehicles. A Eurelectric Position Paper (2016). http://www.eurelectric.org/media/285584/ev_and_charging_infrastructure_final-2016-2310-0001-01-e.pdf
35. Electric Vehicle Charging Infrastructure. http://www.eafo.eu/electric-vehicle-charging-infrastructure

Estimating Pollutant Emissions Based on Speed Profiles at Urban Roundabouts: A Pilot Study

Orazio Giuffrè[1], Anna Granà[1(✉)], Tullio Giuffrè[2], Francesco Acuto[1], and Maria Luisa Tumminello[1]

[1] Department of Engineering, University of Palermo, Palermo, Italy
{orazio.giuffre,anna.grana,francesco.acuto,
marialuisa.tumminello01}@unipa.it
[2] University of Enna Kore, Enna, Italy
tullio.giuffre@unikore.it

Abstract. The paper describes the pilot study conducted to assess the feasibility of the empirical approach utilizing vehicle trajectory data from a smartphone app and the Vehicle-Specific Power methodology to estimate pollutant emissions at urban roundabouts. The goal of this research phase is to acquire instantaneous speed data from a sample of six roundabouts located in the road network of the City of Palermo, Italy, and quantify emissions generated by the test vehicle through the examined roundabouts. For the case studies of roundabouts acceleration events in the circulating and exiting areas contributed to about 25% of the emissions for a given speed profile. More in general, the results from this research shed lights for further opportunities to examine infrastructural scenarios when decision makers require to assess changes in the design or operation of urban transportation systems.

Keywords: Roundabout · Pollutant emissions · VSP methodology · Smartphone app

1 Introduction

Land transport sector is facing major changes clearly inspired by the need to contain within certain safety margins problems coming from traffic congestion, road crashes, energy consumption, carbon emissions and other forms of traffic-related air pollution. In this view, the concept of sustainable mobility provided the basis for the discussions which led to invest resources so that these problems are minimized, when not even zeroed internationally. In 2011 the European Commission adopted specific transport policy objectives with the White Paper "Towards a competitive and resource efficient transport system" [1], in order to combine the increase of mobility with the reduction of traffic emissions through a wide-ranging strategy and with a long-time horizon. The target is to achieve a 60% reduction in greenhouse gas emissions (GHG) by 2050 compared to 1990 levels. This objective would be the contribution offered by the transport sector to the comprehensive EU objective of an 80%–95% GHG reduction as foreseen in the Roadmap for a low carbon economy.

Road transport sector is one of the main sources of emissions to air. Automotive companies are under pressure for a long time to reduce the toxic levels of vehicle exhaust emissions. Hybrid-electric vehicles cut pollution and comply with laws limiting vehicle exhaust emissions, but electric cars, which are already available on the market today, are a very interesting option compared to diesel, petrol, LPG or CNG. In addition to being a good choice for the environment, electric vehicles also began to be so with regard to performance and running costs. Although emissions of air pollutants from road transport have significantly decreased in recent years thanks to technological innovations, the adaptation of the vehicle fleet to the emission standards of the new vehicles proceeds on the basis of the physiological rate of replacement of the vehicle fleet. Furthermore, the average age of the vehicular fleet is quite high as the penetration rate of modern technologies is slow.

Emission models can predict emissions at regional or national level, permitting to obtain emission inventories at these levels or they can predict the effects on emissions produced by changes in the design or operation of urban transportation systems at a very local level. In literature a further distinction is between the average speed approach and the instantaneous speed approach. Estimation of exhaust pollutants emissions produced by a car engine is still an important applied research topic because of the health effects and impacts on the environment. Vehicle emissions are affected by various factors as the acceleration and deceleration of vehicles, signals, and idle time [2–4]. Measurement of road vehicle exhaust emissions are some of the key inputs to air pollution models which are used to estimate air pollution levels, human exposure to traffic-related air pollution, and associated health effects.

Based on the above the paper describes the pilot study conducted to explore and assess the feasibility of the empirical approach utilizing instantaneous speed data from a smartphone app and the vehicle-specific power (VSP) methodology based on onboard measurements in light passenger vehicles [5]. The main goal of this research is to acquire vehicle trajectory data from a sample of six roundabouts located in the road network of the City of Palermo, Italy, and quantify emissions generated by the test vehicle through the examined roundabouts.

Before describing data collection for the roundabout sample, which included geometric features and speed profiles, and the instantaneous emission estimation, a literature review on different types of vehicular emission models is introduced. The main findings and recommendations for future developments of the research are explained in the conclusion section, together with the limitations of this study.

2 Summary of Literature on Emission Models

Models for estimating pollutant emissions from road traffic can be average speed-based models and dynamic instantaneous emission models.

According to [6] the average speed-based approach estimates emissions by using information aggregated by vehicle type as derived from driving patterns. The principle behind the average speed-based models considers that the average emission factor for a certain pollutant and a given type of vehicle varies with the average speed during a move; such models are used with macroscopic traffic flow models. The average speed models widely used in Europe and US use the emissions factors of the COPERT and TRL [7, 8]. In spite of their conceptual simplicity in calculating emissions with few input data (average speed, traffic volume, and link length), limitations are the inability to capture the speed variation in the case of acceleration, deceleration, and idling [9], and to explain the ranges of vehicle operation and emission behavior at a given average speed [10]. Thus, average speed-based models can underestimate emissions especially in urban settings. Some authors [11, 12] improved accuracy of average speed models by using real-time data, but more research should be done to account for various vehicle types and to get better the predictive performance of these models.

Dynamic instantaneous emission models or fuel consumption models estimate instantaneous vehicle fuel consumption and second-by-second vehicle emission rates based on the instantaneous speed and acceleration of individual vehicles [13]. The instantaneous speed approach uses the acceleration rate (or the product speed-acceleration) in addition to speed. Emission functions can be generated by assigning defined emission values to particular operational conditions.

Instantaneous emission models are used with microscopic traffic flow models to produce accurate estimates of emissions and fuel consumption [9], but they need large computation times for large-scale networks. However, these models result sensitive to changes in vehicle acceleration and can be used in the evaluation of operational level transportation projects such as roundabouts [14]. CMEM [15] is one of the representative instantaneous emission models that predicts second-by-second fuel consumption, and tailpipe emissions of CO, CO_2, HC, and NO_x based on different modal operations from in-use vehicle fleet. In this model, the entire fuel consumption and emissions process is broken down into components corresponding to physical phenomena associated with vehicle operation and emissions production. CMEM calculates emissions by utilizing each vehicle's driving cycle data, but it is not appropriate for local roads because it is impossible to collect data on every vehicle in traffic [12]. Another widely used instantaneous speed approach estimates emissions by using a modelling technique featuring real-time engine power (Vehicle Specific Power, in the following VSP) [5]; such approach is used to assess the impact of vehicle operating conditions on emissions and energy consumption. The VSP estimates depend on the speed, roadway grade and acceleration or deceleration on the basis of the second-by-second cycles. VSP captures dependence of emissions on power, includes the impact of different levels of accelerations and speed changes on emissions, and accounts for the effect of road infrastructure on power demand [16].

Second-by-second Global Positioning System (GPS) data give flexibility to characterize vehicle emissions by using VSP under real-world conditions at any location [17]. VSP has been incorporated into vehicle emission models as MOVES [5, 18, 19]. Among the studies that have applied VSP methodology, one can remember Yao et al. [20] that examined the role of freeway grade in VSP profiling; their results demonstrated that the sample distribution of VSP gives a better fit at lower grades. However, this study provided just a reference for preparing operating mode distribution inputs for MOVES model, since grade-specific VSP distributions for arterials and locals roads should be studied in more depth. Song et al. [3] estimated VSP distribution among urban restricted access roadways; the study suggested that the distribution of VSP at various speed bins follows the normal distribution. Zhao et al. [21] confirmed that the normal distribution is best suited when travel speed of light duty vehicles and heavy duty vehicles moving along freeways is lower than 90 km/h. Liao et al. [22] proposed a simulation model for a signalized intersection where light duty vehicles were equipped with a cooperative vehicle-infrastructure system; the results confirmed that the environmental benefits depended on drivers' compliance behaviors.

In the latest years there have been many developments on this field, but few applications concerned roundabouts. Coelho et al. [23] found three speed profiles covering all combinations of stop and no-stop conditions for vehicles entering a single-lane roundabout, and developed a methodology to quantify the emission impact of the operational performance related to stop-and-go behavior. They used the VSP approach, identified the parameters associated with the occurrence of changes in speed cycles with influence on emissions and explained the interaction among operation variables, geometry and the resulting traffic emissions.

Results from studies reported in literature showed how roundabouts located on urban corridors affected traffic performance and pollutant emissions from vehicles. Salamati et al. [24] based on [23] tried to understand the difference between the characteristics of pollutant emissions at multi-lane and single-lane roundabouts; emissions were measured by using VSP methodology. The main findings from the research highlighted that differences in emission estimation between left and right lane movements should be redefined with greater depth and scope. In turn, Fernandes et al. [25] finalized their research to identify the hotspots along the corridors where emissions tend to be consistently high and found no significant differences between emissions of roundabouts characterized by similar geometric design and fairly spaced along the corridor. However, other corridors with different spacing among roundabouts in sequence should be examined to better characterize the spatial distribution of emissions and assess the impact of entry deflection angle on acceleration profiles.

Further researches compared the emission performances of conventional and innovative roundabouts by means of VSP methodology to estimate second-by-second pollutant emissions for a mixed fleet of conventional vehicles also in urban settings [14, 26]. However, more experimental data should be gathered and analyzed to obtain generalizable conclusions.

Further studies have estimated emissions from traffic by integrating traffic microscopic simulation models and external emissions models. In this regard Stogios et al. [27] determined the effects of driving settings with Automated Vehicles (AVs) on Greenhouse Gas (GHG) emissions at an urban corridor by traffic microsimulation and emissions modelling. No significant impact on GHG emission reductions was obtained when driving settings included AVs alone; in turn, the inclusion of vehicle powertrain technology has potential for a maximum of 24% in GHG emission reduction. However, given that the specific topic is outside the objectives of this research phase, the reader is referred to specialistic literature.

It should be noted that VSP methodology has few applications to roundabout case studies until today. Thus, the paper starts from a reflection on the gap in the current literature with regard to the assessment of pollutant emissions at existing roundabouts. The overview of models for estimating emissions from vehicles in Table 1 is only provided for illustrative purposes.

Table 1. Models for estimating emissions from vehicles.

Model	Scale	Basic variable	Input data	Typical application
COPERT [7, 8]	Macroscopic	Average speed-based	Average trip speed	Regional or national emission inventories, dispersion modelling
EMFAC [28]	Macroscopic	Trip-based vehicle average speed	Vehicle miles traveled, emission rates	Emissions inventories, impact on local roadways and intersections
TREM [29]	Macroscopic	Link-based vehicle average speed	Traffic volume, vehicle speed, vehicle distribution by categories and by classes, road length	Emissions inventories regional-, local- and city-level, dispersion modelling
aaSIDRA [30]	Mesoscopic	Four-modal activity	Vehicle parameters, speed, acceleration rate, grade, cost parameters	Assessment of environmental impacts, cost, energy and air pollution implications of intersection design
CMEM [15]	Microscopic	Instantaneous speed	Individual vehicle variables (speed, acceleration, road grade) and fleet composition	Regional inventories, emissions benefits of project-level or corridor-specific control measures, ITS implementations
MOVES [19]	Microscopic	Vehicle specific power (VSP) and instantaneous speed	Vehicle operating time, emission rates	Multiple scale analysis, emission inventories
MODEM[a] [31]	Microscopic	Instantaneous speed	Driving pattern	Temporal and spatial analysis of emissions, dispersion modelling

[a] Type of emission factor/function: discrete, trip-based.

3 Data and Research Methodology

3.1 Data Collection

A set of six four-legged roundabouts located in road network of the City of Palermo, Italy, was selected for this study; the layouts are shown in Fig. 1, while Table 2 shows some characteristics of the roundabout case studies.

The research team collected data at the roundabouts during the morning (7:00–8:30 a.m.) peak periods on regular weekdays (Wednesday to Thursday) in October and November, 2018. The speed limit in the studied roundabouts is 50 km/h. Vehicle trajectory data were collected by using the app Speedometer GPS PRO for Android smartphone which calculates the second-by-second detailed GPS trajectory. Location, travel time, distance, grade, second-by-second instantaneous speed and acceleration are the values directly extracted from the GPS data. The left turn and through movements at multi-lane sites were experienced by entering each roundabout from the left entry lane; from 7 to 10 runs were done for a total of 236 GPS travel runs (188 travel runs of through movements in both directions and 48 travel runs of left turn movements).

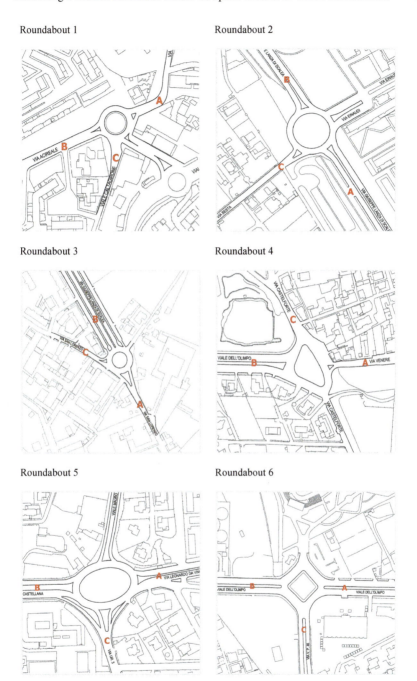

Fig. 1. Layouts of the surveyed roundabouts of the pilot study. Note: A, B, C only denote the legs interested by through movements (A to B and viceversa) and left turns.

Table 2. Information on the roundabout case studies.

Roundabout	Outer diameter [m]	Circulating width [m]	Entry (Exit) roadway width[a] [m]	Total entry flow [vph]	Approach speed[b] [km/h]	Circulating speed [km/h]
Roundabout 1	48.00	7.00	7.00 (6.00)	1576	25	18
Roundabout 2	40.00	8.00	9.00 (9.00[c])	3984	25	20
Roundabout 3	30.00	10.00	7.00 (7.00)	2336	23	20
Roundabout 4	40.00	11.00	10.00[d] (9.00)	1312	30	30
Roundabout 5	45.00	9.00	8.00 (10.50[e])	4052	20	30
Roundabout 6	40.00	10.00	10.00 (10.00[f])	988	30	25

Note: [a]Effective entry width before widening the entry approach or before by-pass for right turners; [b]average value of speed at the entry line; [c]5.00 m for the one-lane entry and 4.00 m for the one-lane exit before widening (Besta Street); [d]3.50 m for the one-lane entry (exit) (Castelforte Street); [e]4.50 m for the one-lane exit (Sarullo Street); [f]7.00 m for exit in Pertini Street

The instantaneous speed and acceleration-deceleration profiles were estimated from data on vehicle dynamics using a Light passenger diesel vehicle (LPDV) conforming to Euro IV Emission Standard as the test vehicle. The characteristics of the vehicle are within the tested LPDV specifications used to obtain the emissions factors for VSP methodology; see [32]. Entry and conflicting traffic flows were also videotaped in some sites; they were not reported here in a complete way since useless for the purpose of this research step. Note that the examined roundabouts are sited in areas different from the urbanistic point of view, and the percentage of the heavy vehicles didn't overcome 10% almost during the observational time periods.

Approximately 90 km of road were travelled. Based on [14, 33] the number of runs per site was deemed appropriate to obtain appropriate results from the collected data; over 15 h were gathered for the surveyed roundabouts. Each speed profile that occurred for the vehicle entering the roundabout was related to the greater or lesser congestion level for conflicting and entry flows. The speed profiles were consistent to what experienced by [23] for vehicles approaching the roundabouts; see Figs. 2, 3 and 4 for some speed profiles obtained from processing GPS data collected thorough the examined roundabouts. These speed trajectories were later used to estimate the emissions generated from the roundabouts. The test vehicle experienced one stop at the entry line before finding a useful headway, accelerated to move along the ring and to exit the roundabout (see Fig. 2); the test vehicle also experienced entry and negotiation of the circulating area without stopping, and then accelerating back to cruise speed as it is exiting (see Fig. 3). Some cases of multiple stopping were also experienced depending on the level of congestion of the approach (see Fig. 4). In some cases, the vehicle waited for a useful headway for more time, thus entered the roundabout facing low circulating traffic, or the vehicle spent the time in deceleration as it approached the roundabout, entered the circulating lanes at low speed and acceleration as it exited the roundabout. In other cases, especially for left turning manoeuvres, the test vehicle experienced entry and negotiation of the circulating area without stopping, and then accelerating back to cruise speed as it exited. By way of example, the speed profile in Fig. 4 shows that the vehicle travelling in roundabout faced with higher idle and low speed situations at the downstream and circulating areas compared to less congested traffic situations.

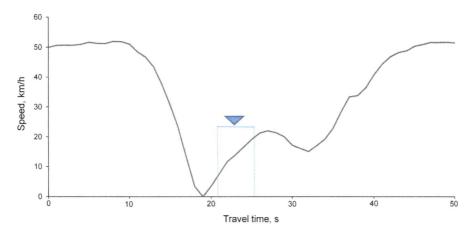

Fig. 2. Speed profile for through movement at Castelforte roundabout (Roundabout 4, A-B movement)

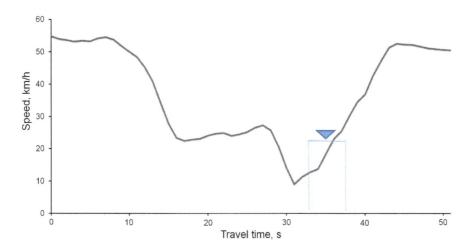

Fig. 3. Speed profile for left turn movement at Pertini roundabout (Roundabout 6, B-C movement)

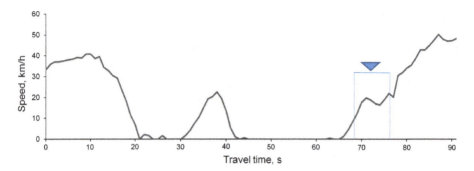

Fig. 4. Speed profile for through movement at Castellana roundabout (Roundabout 5, A-B movement)

3.2 Instantaneous Emission Estimation

Vehicle emissions strongly depend on specific power of the vehicle. Differently from average speed-based emission models [12, 34], instantaneous emission models relate emission rates to second-by-second vehicle operations, and consider the dynamics of driving cycles to estimate emissions for any vehicle operation profile [10]. In this pilot study, a VSP-based emission model was applied to calculate the instantaneous emissions of the test vehicle used to collect the speed per second-by-second, acceleration or deceleration, and road grade data at the roundabouts under examination [16].

Vehicle Specific Power (VSP) is conventionally defined as the instantaneous power per unit mass of the vehicle. The VSP expression was developed by [5]; according to [5] the instantaneous power generated by the engine is applied to overcome the rolling and aerodynamic resistances and to increase both the kinetic and potential vehicle energy. The VSP values are put into 14 modes of engine regime, and emission factor values for each mode are used to estimate CO_2, CO, NO_x and HC emissions. The VSP values for light duty vehicles are calculated by Eq. 1 [14, 18]:

$$VSP = v \cdot [1.1 \cdot a + 9.81 \cdot sin(arctan(grade)) + 0.132] + 0.000302 \cdot v^3 \qquad (1)$$

where VSP is the Vehicle Specific Power [kW/ton]; v is the vehicle instantaneous speed [m/s]; a is the vehicle acceleration or deceleration rate [m/s^2]; grade is the terrain gradient as decimal fraction.

Table 3 shows the average values for CO_2, CO, NO_x, and HC emission rates [g/s] for VSP modes for Light Passenger Diesel Vehicles (LPDV) as referred by [32].

The average values of emissions rates by VSP mode for Light Passenger Gasoline Vehicles (LPGV) and Light Commercial Diesel Vehicles (LCGV) can be also found in [14]. Fernandes et al. [14] refer that the average emission rates for pollutants CO_2, CO, NO_x and HC by VSP mode are the average of tailpipe emission measures collected from over 40 light passenger vehicles using a Portable Emissions Measurement System at North Carolina State University (NCSU).

Table 3. Average values of CO_2, CO, NO_x, and HC emissions rates by VSP mode for LPDV.

VSP range [Kw/ton]	VSP mode	Average modal emission rates [g/s]			
		CO_2	CO	NO_x	HC
VSP < -2	1	0.21	0.00003	0.0013	0.00014
-2 ≤ VSP < 0	2	0.61	0.00007	0.0026	0.00011
0 ≤ VSP < 1	3	0.73	0.00014	0.0034	0.00011
1 ≤ VSP < 4	4	1.50	0.00025	0.0061	0.00017
4 ≤ VSP < 7	5	2.34	0.00029	0.0094	0.00020
7 ≤ VSP < 10	6	3.29	0.00069	0.0125	0.00023
10 ≤ VSP < 13	7	4.20	0.00058	0.0155	0.00024
13 ≤ VSP < 16	8	4.94	0.00064	0.0178	0.00023
16 ≤ VSP < 19	9	5.57	0.00061	0.0213	0.00024
19 ≤ VSP < 23	10	6.26	0.00101	0.0325	0.00028
23 ≤ VSP < 28	11	7.40	0.00115	0.0558	0.00037
28 ≤ VSP < 33	12	8.39	0.00096	0.0743	0.00042
33 ≤ VSP < 39	13	9.41	0.00077	0.1042	0.00040
VSP ≥ 39	14	10.48	0.00073	0.1459	0.00042

For the particular test vehicle, the emissions values of CO_2, CO, NO_x and HC pollutants were estimated from the distribution of time spent in each VSP mode obtained from the speed profiles, according to Eq. 2 as found in [14]:

$$E_{ij} = \sum_{k=1}^{N_k} F_{kj} \qquad (2)$$

where E_{ij} = total emissions for source pollutant j and speed profile i, [g]; k is the label for second of travel [s]; F_{kj} is the emission factor for pollutant j in label for second of travel k [g/s]; N_k is the number of seconds [s].

In order to estimate the pollutant emissions for a given speed profile i, Eq. 1 is used to calculate second-by-second emission rates for the vehicle test which experienced the speed profile i. An average influence area of 500 m was established to maintain consistency among runs and calculate the pollutant emissions for each speed profile at the examined roundabouts. It was defined as the sum of the deceleration distance of a vehicle travelling from the cruise speed as it approaches and the enters the roundabout, and the acceleration distance as it exits the roundabout up to the section it reaches the cruise speed. Note that each case study of roundabout is characterized by grades less than 2% so that this parameter was neglected. Figure 5 shows the relative frequencies of time spent in each VSP mode for the speed profiles in Figs. 2 and 4. In the first case (see Fig. 5a) the test vehicle spent most of the time in VSP modes 4 and 5 corresponding to accelerations as it exited the roundabout. In the case of Fig. 5b the vehicle spent most of time in VSP modes 2 (deceleration), 3 (idle) and 4 (acceleration and cruising). The percent of the time spent in VSP modes higher than 5 is low in both case.

Total emission by each pollutant and roundabout was obtained as average of emission for source pollutant and speed profile.

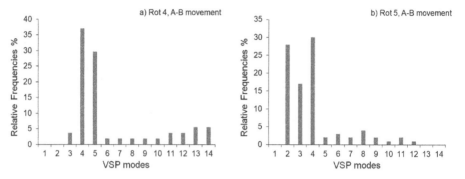

Fig. 5. Relative frequencies of time spent in VSP modes

4 Results

This section addresses the results of the pollutant emission impact from characteristic speed trajectories and VSP methodology. Based on emission rates corresponding to each run and each (through and left turn) movement, average CO_2, CO, NOx and HC emission rates were estimated for the roundabouts in Fig. 1.

Total emission [g] for the different pollutants were estimated and compared among the examined roundabouts; no particular critical issues were identified with respect to the EURO IV standards on pollutant emissions from vehicular sources. The results were also consistent with the expectation of lower CO and higher NOx emissions than gasoline vehicles as found by [32].

Figure 6 shows, by way of example, the CO_2 total emissions for through and left turn movements as experienced by the test vehicle in this pilot study and the corresponding root mean squared error (less than about 5%). It is well known that several pollutants have been phased out since 1992 with the introduction of the legislation on this topic; CO_2 is not really toxic but is regulated because it increases the greenhouse effect. In accordance with the EU's air quality directives, the CO_2 emission target for the test vehicle corresponds to 130 g/km; thus, for the cases examined with the test vehicle (see Fig. 3), the CO_2 emissions were found to be higher than expected.

By way of example, the NOx total emissions and the (HC + NOx) total emissions for through movements as experienced by the test vehicle at the roundabout sample are also shown in Fig. 7. In accordance with the EU's air quality directives, the NOx emission target is corresponding to 0.25 g/km, while the (HC + NOx) emission target is corresponding to 0.30 g/km. Major differences of emissions among roundabouts were attributable to the different acceleration conditions that the test vehicle experienced during the runs.

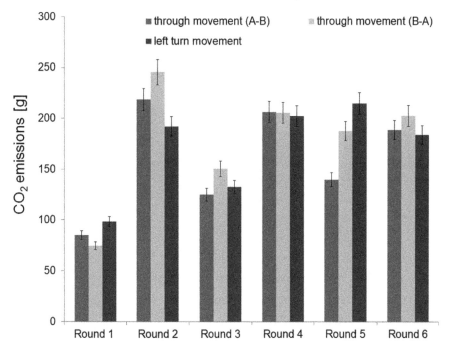

Fig. 6. CO_2 total emission [g] for the pilot sample of roundabouts (see Fig. 1 for movements) Round stands for Roundabout. Note that left turn movements are corresponding to the following: Round 1 (A-C); Round 2 (A-C); Round 3 (A-C); Round 4 (B-C); Round 5 (C-B); Round 6 (B-C)

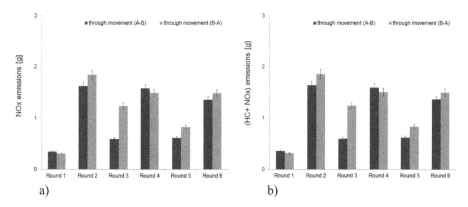

Fig. 7. Total emission [g] for the pilot sample of roundabouts (see Fig. 1 for movements): (a) NOx emissions; (b) (HC+NOx) emissions. Note that Round stands for Roundabout.

Figure 8 shows the cumulative distributions of CO_2 emissions with reference to the influence area of two multi-lane roundabout of the pilot sample in Fig. 1; in particular, Fig. 8a shows the spatial distribution of CO_2 emissions for the speed profile in Fig. 2 (Roundabout 4, A-B movement), while Fig. 8b shows the spatial distribution of CO_2 emissions for the speed profile in Fig. 4 (Roundabout 5, A-B movement).

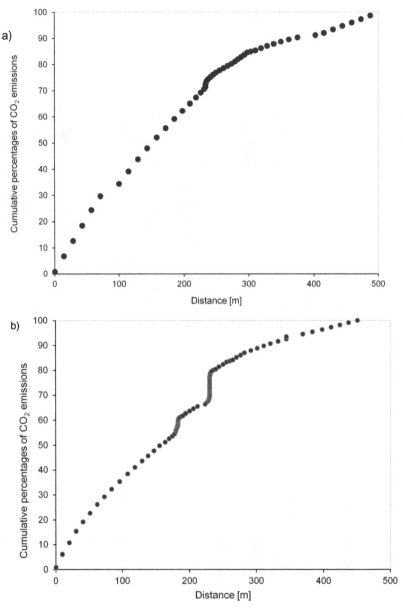

Fig. 8. Spatial distribution of pollutant emissions: (a) Roundabout 4, A-B movement; (b) Roundabout 5, A-B movement.

Similar patterns were exhibited for the analogous distributions of CO_2 emissions for the other roundabouts. The graphs in Fig. 8a shows that the vehicle experienced a short stop-and-go and about the 25% of CO_2 total emission is produced during the acceleration from zero to the cruise speed. In turn, Fig. 8b shows that the relative increase of emissions percentage with the distance is higher in the acceleration mode and during long stop-and-go rather than during short stop-and-go. According to [14, 35] acceleration events in the circulating and exiting areas of a roundabout contributed to more than 25% of the emissions for a given speed profile.

5 Conclusions

The paper describes the pilot study conducted to assess pollutant emission estimations by means of an empirical approach using instantaneous speed data from a smartphone app and the vehicle-specific power (VSP) methodology. This step of research is done to acquire vehicle trajectory data from a sample of roundabouts located in road network of the City of Palermo, Italy, and to quantify emissions generated by a test vehicle.

Vehicle trajectory data were collected by using the app Speedometer GPS PRO for Android smartphone which calculated the second-by-second detailed GPS trajectories. The left turn and through movements were experienced entering each roundabout for a total of 236 travel runs of through movements in both directions and left turn movements; over 15 h were gathered for the surveyed roundabouts.

Each speed profile that occurred for the vehicle entering the roundabout was related to the greater or lesser congestion level for conflicting and entry flows.

Consistently with studies in literature on the topic (e.g. [14, 23, 35]) the results highlighted that acceleration events in the circulating and exiting areas of roundabouts produced around 25% of the emissions for a given speed profile.

With the trajectory data collected from a smartphone app, speed and acceleration (deceleration) were obtained directly. Thus, it was possible to explore the driving performance at the roundabouts in Fig. 1 from an environmental point of view.

The results show the feasibility of the approach that used vehicle trajectory data from a smartphone app and the Vehicle-Specific Power methodology to estimate pollutant emissions at urban roundabouts. The approach is revealed friendly both in data collection and in the following data analysis. In this sense, it proves to be an opportunity to gather data on a wide scale through a smart community.

It should be noted that the results have to be seen in relation to the number of selected case studies and the test vehicle used to collect on-field data. Given the need for new solutions supporting data collection and analysis of environmental impact in road transport, the results from this pilot study necessarily lead to further research to identify behaviors and road performance not limited to the application of VSP methodology, and environment considerations.

Since a transportation engineer must consider the trade-off between different factors as traffic emissions and road safety, future developments can include: gathering data on a great number of roundabouts, particularly, in those with greater variability in terms of approach, entry or circulating speeds, geometric features, and traffic flows; assessment of further infrastructural and traffic scenarios through microsimulation and development of models to relate the speed profile to the entry capacity, simulation of traffic operations in presence of AVs and evaluation of their impact on reduction of polluting emissions, and so on. However, taking account of the above considerations and given the need to ensure smart tools supporting the reduction of environmental impacts in road transport, the study can represent the starting point for future assessment of infrastructural options when decision makers require to assess changes in the design or operation of urban transportation systems in support of their choices.

References

1. WHITE PAPER Roadmap to a Single European Transport Area – Towards a competitive and resource efficient transport system (2011). https://eur-lex.europa.eu/legal-content/EN/ALL/?uri=CELEX:52011DC0144
2. Park, S., Lee, J., Lee, C.: State-of-the-art automobile emissions models and applications in North America and Europe for sustainable transportation. KSCE J. Civ. Eng. **20**(3), 1053–1065 (2016)
3. Song, G., Yu, L., Tu, Z.: Distribution characteristics of vehicle-specific power on urban restricted-access roadways. J. Transp. Eng. **138**(2), 202–209 (2012)
4. Giuffre, O., Grana, A., Giuffre, T., Marino, M.: Emission factors related to vehicle modal activity. Int. J. Sustain. Dev. Plan. **6**(4), 447–458 (2011)
5. Jiménez-Palacios, J.: Understanding and quantifying motor vehicle emissions with vehicle specific power and TILDAS remote sensing. Ph.D. dissertation. Massachusetts Institute of Technology, Cambridge. Mass. (1999)
6. Achour, H., Carton, J.G., Olabi, A.G.: Estimating vehicle emissions from road transport, case study: Dublin City. Appl. Energy **88**, 1957–1964 (2011)
7. Katsis, P., Ntziachristos, G.M.: Description of new elements in COPERT 4 v10.0 (12. RE.012.V1). EMISIA SA, Thessaloniki, Greece (2012)
8. Ntziachristos, L., Samaras, Z.: COPERT III: computer program to calculate emissions from road transport – Methodology and emissions factors (version 2.1). Technical report 49. European Environment Agency (2000)
9. Ahn, K., Rakha, H.: The effects of route choice decisions on vehicle energy consumption and emissions. Transp. Res. Part D **13**(3), 151–167 (2008)
10. Boulter, P.G., McCrae, I.S., Barlow, T.J.: A review of instantaneous emission models for road vehicles. Published Project Report 267. TRL (2007). https://docs.niwa.co.nz/library/public/PPR267.pdf
11. Negrenti, E.: The 'Corrected Average Speed' approach in ENEA's TEE model: an innovative solution for the evaluation of the energetic and environmental impacts of urban transport policies. Sci. Total Environ. **235**, 411–413 (1999)
12. Ryu, B.Y., Jung, H.J., Bae, S.H.: Development of a corrected average speed model for calculating carbon dioxide emissions per link unit on urban roads. Transp. Res. Part D **34**, 245–254 (2015)
13. Ajtay, D., Weilenmann, M.: Static and dynamic instantaneous emission modelling. Int. J. Environ. Pollut. **22**(3), 226–239 (2004)
14. Fernandes, P., Pereira, S.R., Bandeira, J.M., Vasconcelos, L., Bastos Silva, A., Coelho, M.C.: Driving around turbo-roundabouts vs. conventional roundabouts: are there advantages regarding pollutant emissions? Int. J. Sustain. Transp. **10**(9), 847–860 (2016)
15. CMEM National Academies of Sciences, Engineering, and Medicine. Predicting Air Quality Effects of Traffic-Flow Improvements: Final Report and User's Guide. The National Academies Press, Washington, DC (2005). https://doi.org/10.17226/13797
16. Kutz, M.: Environmentally Conscious Transportation. Wiley, New York (2008)

17. Frey, H.C., Zhang, K., Rouphail, N.M.: Fuel use and emissions comparisons for alternative routes. Time of day. Road grade and vehicles based on in-use measurements. Environ. Sci. Technol. **42**(7), 2483–2489 (2008)
18. USEPA. Methodology for developing modal emission rates for EPA's multiscale motor vehicle and equipment emission system (EPA420). North Carolina State University for USEPA, Ann Arbor, MI (2002)
19. Sontag, D., Gao, H.O.: The MOVES from MOBILE: preliminary comparisons of EPA's current and future mobile emissions models. In: TRB 86th Annual Meeting, Washington D. C., USA, 21–25 January 2007
20. Yao, Z., Wei, H., Liu, H., Li, Z.: Statistical Vehicle Specific Power Profiling for Urban Freeways. Procedia Soc. Behav. Sci. **96**, 2927–2938 (2013)
21. Zhao, Q., Yu, L., Song, G.: Characteristics of VSP distribution of light-duty and heavy-duty vehicle on freeway. J. Transp. Syst. Eng. Inf. Technol. **15**(3), 196–203 (2015)
22. Liao, R., Chen, X., Yu, L., Sun, X.: Analysis of emission effects related to drivers' compliance rates for cooperative vehicle-infrastructure system at signalized intersections. Int. J. Environ. Res. Public Health **15**(1), pii:E122 (2018)
23. Coelho, M.C., Farias, T.L., Rouphail, N.M.: Effect of roundabout operations on pollutant emissions. Transp. Res. Part D Transp. Environ. **11**(5), 333–343 (2006)
24. Salamati, K., Coelho, M., Fernandes, P., Rouphail, N.M., Frey, H.C., Bandeira, J.: Emission estimation at multilane roundabouts: effect of movement and approach lane. Transp. Res. Rec. **2389**, 12–21 (2014)
25. Fernandes, P., Salamati, K., Rouphail, N., Coelho, M.: Identification of emission hotspots in roundabouts corridors. Transp. Res. Part D Transp. Environ. **37**, 48–64 (2015)
26. Fernandes, P., Coelho, M.: Making compact two-lane roundabouts effective for vulnerable road users: an assessment of transport-related externalities. In: Macioszek, E., Akçelik, R., Sierpiński, G. (eds.) Roundabouts as Safe and Modern Solutions in Transport Networks and Systems, vol. 52, pp. 99–111. Springer, Heidelberg (2018)
27. Stogios, C., Saleh, M., Ganji, A., Tu, R., Xu, J., Roorda, M.J., Hatzopoulou, M.: Determining the effects of automated vehicle driving behavior on vehicle emissions and performance of an urban corridor. In: Transportation Research Board 97th Annual Meeting, Washington DC, United States, 7–11 January 2018
28. EMFAC2014 Version 1.0.7 User Guide. California Air Resource Board, Sacramento, California, April 2014. https://www.arb.ca.gov/msei/downloads/emfac2014/emfac2014-vol1-users-guide-052015.pdf
29. Transport Emission Model for Line Sources (TREM) – Methodology- Technical report EIE/07/239/SI2.466287. Department of Environment and Planning, University of Aveiro. https://ec.europa.eu/energy/intelligent/projects/sites/iee-projects/files/projects/documents/t.at._trem_methodology_en.pdf
30. Akçelik and Associates aaSIDRA User Guide, Version 2.0. Akcelik and Associates Pty Ltd., Melbourne, Australia (2002)
31. Jost, P., Hassel, D., Webber, F.-J.: Sonnborn: emission and fuel consumption modelling based on continuous measurements. Deliverable No. 7 DRIVE Project V1053. TUV Rhineland, Cologne (1992)
32. Coelho, M.C., Frey, C.H., Rouphail, N.M., Zhai, H., Pelkmans, L.: Assessing methods for comparing emissions from gasoline and diesel light-duty vehicles based on microscale measurements. Transp. Res. Part D Transp. Environ. **14**, 91–99 (2009)
33. Li, S., Zhu, K., van Gelder, B., Nagle, J., Tuttle, C.: Reconsideration of sample size requirements for field traffic data collection with global positioning system devices. Transp. Res. Rec. **1804**, 17–22 (2002)

34. Weilenmann, M., Soltic, P., Saxer, C., Forss, A.M., Heeb, N.: Regulated and unregulated diesel and cold start emissions at different temperature. Atmos. Environ. **39**, 2433–2441 (2005)
35. Mudgal, A., Hallmark, S., Carriquiry, A., Gkritza, K.: Driving behavior at a roundabout: a hierarchical Bayesian regression analysis. Transp. Res. Part D Transp. Environ. **26**, 20–26 (2014)

Examining the Influence of Railway Track Routing on the Thermal Regime of the Track Substructure – Experimental Monitoring

Peter Dobeš[✉], Libor Ižvolt, and Stanislav Hodás

Faculty of Civil Engineering, University of Žilina, Žilina, Slovakia
{dobes,libor.izvolt,stanislav.hodas}@fstav.uniza.sk

Abstract. Based on the experimental monitoring results, the paper identifies the influence of non-traffic load (water, snow, frost) on the track substructure freezing, depending on the track routing method (embankment, cut).

The paper introduction characterizes two railway track models (scale 1:1), which are a part of the Experimental Stand DRETM, (in the influence assessment the 1st measuring profile - embankment and the 5th measuring profile – cut are applied). The second part of the paper presents the results of the experimental monitoring and subsequently the comparison of both measuring profiles, in regard to the achieved value of the track substructure depth of freezing. The paper conclusion evaluates the influence of the railway track routing on the thermal regime and on the track substructure freezing.

Keywords: Railway track · Thermal regime of track substructure · Air frost index · Depth of freezing of track substructure

1 Introduction

The Department of Railway Engineering and Track Management (DRETM) has been studying the effects of non-traffic load on the thermal regime of the track substructure since 2003. The aim is to revise the design methodology for the track substructure dimensioning for the non-traffic load (water and frost). Outside of Slovakia, this issue is addressed primarily in a number of Nordic countries and also by influential Canadian researchers [1–4]. In Canada, the SoilVision software was developed [5]. This software includes 6 different software products, focusing on various problematic processes (soil deformation, slope stability, groundwater seepage, soil contamination, thermal regime of soils…), occurring in both saturated and unsaturated soils.

In 2012 the Canadian software SoilVision [5], specifically SVHeat [6] was purchased, which significantly improved the quality and validity of monitoring the thermal regime of the track at DRETM. The need for input data in the numerical modeling of the thermal regime of the track required the building of a new 1:1 railway track model at the Veľký Diel campus, named Experimental Stand DRETM.

The final scope of the stand structure was not only affected by the software *Soilvision*, which requires specific inputs to the numerically model the thermal regime but also experience with the previously-built experimental stand. It was built in the premises of the Faculty of Civil Engineering, which was then located outside of the current campus. Measurements carried out on that experimental stand, which only included one measuring profile – track model on the surrounding terrain level, were completed in 2016.

The construction of a new experimental stand was completed in 2017. The Experimental Stand DRETM at present consists of 6 measuring profiles, which differ in the type of the structural arrangement and earthwork shape. It enables to monitor the climatic factors as well as the deformation characteristics of the materials applied in the track substructure. The stand layout can be observed in Fig. 1.

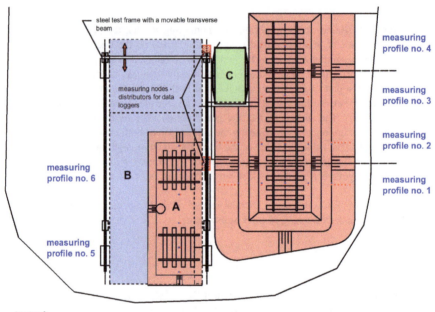

Legend:

Sector **A**: verification of the material thermotechnical characteristics and monitoring of the thermal regime of the track substructure construction
Sector **B**: measuring of the deformation characteristics - the major experimental stand
Sector **C**: measuring of the deformation characteristics - the small experimental stand

Fig. 1. Experimental Stand DRETM – layout (source: own research)

The influence of the track routing (earthwork in an embankment or in a cut) on the track substructure freezing was verified on the measuring profile no. 1 - embankment and measuring profile no. 5 - cut. The measurement profile no. 1 was built in 2013, with its track substructure consisting of a clay subgrade with the addition of river gravel, an embankment of crushed aggregate fr. 0/63 mm, protective layer of crushed aggregate fr. 0/31.5 mm, 450 mm thick, and the ballast bed of gravel fr. 31.5/63 mm, 550 mm thick (Fig. 2).

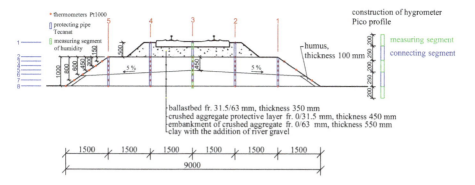

Fig. 2. Experimental Stand – measuring profile no. 1 (source: own research)

The measuring profile no. 5 was completed in 2016. Its track substructure consists of clay subgrade with the addition of river gravel, separation geotextile, protective layer of crushed aggregate fr. 0/31.5 mm, 450 mm thick, and a ballast bed of the identical thickness and fraction as the measuring profile no. 1 (Fig. 3).

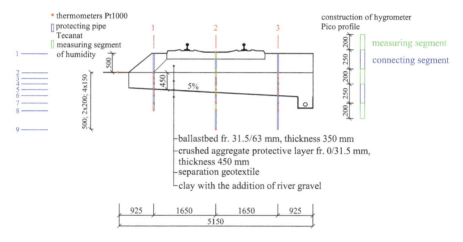

Fig. 3. Experimental Stand – measuring profile no. 5 (source: own research)

The new experimental stand enables to monitor the thermal regime of the railway track using temperature sensors Pt 1000 (red points in Figs. 2 and 3). Moreover, it also allows to monitor the moisture of the applied building materials (water regime) in various depths of measuring profiles, in the individual locations of protective plastic tubes (blue lines in Figs. 2 and 3), with the application of the time-domain reflectometry method (TDR) [7].

The overview of the measurement results, their evaluation and the analysis of experimental monitoring of the thermal regime at the Experimental Stand DRETM for the measuring profiles no. 1 and no. 5 are stated in the following parts of the paper.

2 Verification of the Influence of the Track Routing on the Thermal Regime of the Track

In order to obtain a realistic image of the railway track structure (substructure) freezing, the influence of various factors influencing the penetration of the zero isotherm into the track structure (substructure) is monitored on the Experimental Stand DRETM. The main task of this process is to determine the boundary conditions and limit values of the variables that will be used for numerical modeling in the *Soilvision* software.

The subsequent comparison of the results of the experimental measurements and the results obtained by the numerical modeling of the thermal regime of the railway track will be applied in revising the design of the track substructure dimensioning for the non-traffic load. The updated dimensioning design will meet the needs of the Slovak Railways Company (ŽSR) and will be usable in the design activities of the Slovak railway network modernisation.

One of the many factors that have been verified for the above-mentioned objective is the influence of the track routing (embankment, cut).

For the construction of the railway track in the embankment, the effects of frost from the top and from the sides are assumed, but in the case of a cut, only the frost effects from the top are logically considered.

2.1 Experimental Monitoring of the Thermal Regime of the Track Substructure Carried Out on the Experimental Stand DRETM

The most significant characteristics of the experimental monitoring evaluation of the thermal regime of the track structure in winter is the determination of the position of the zero isotherm in the track substructure (depth of the track substructure freezing D_F). It depends on several factors (average annual air temperature, intensity, length and course of the frost periods, thermal technical parameters of building materials of the track substructure).

The thermal regime of the railway track structure (track substructure) in each measuring profile of the Experimental Stand DRETM can be identified by the temperature sensors embedded in the individual structural layers and their boundaries. A total of 150 temperature sensors Pt 1000 are built in the entire experimental stand structure (6 measuring profiles) and the temperature recording interval is every 30 min. Air temperature is recorded at the same time interval using a protected temperature transmitter Comet T3419 located at 2 m above the surrounding terrain.

Among the monitored climatic characteristics, besides the above-mentioned depth of freezing D_F, there are:

- maximum mean daily air temperature during the winter period $\theta_{s,max}$,
- minimum mean daily air temperature during the winter period $\theta_{s,min}$,
- mean daily air temperature θ_s, average annual air temperature θ_m,
- air frost index I_F and surface air frost index I_{FS}.

The relevant climatic characteristics obtained from the evaluation of the individual winter seasons 2013 to 2019 for the measuring profiles no. 1 and no. 5 are demonstrated in Tables 1 and 2.

Table 1. Climatic characteristics based on the experimental stand measurements – measuring profile no. 1.

Winter period	$\theta_{s,max}$ (°C)	$\theta_{s,min}$ (°C)	θ_m (°C)	I_F (°C, day)	I_{FS} (°C, day)	D_F (m)
2013/2014	10.45	−11.7	9.6	−38	−22	0.41
2014/2015	8.5	−10.8	10.2	−77	−32	0.41
2015/2016	5.5	−10.2	9.9	−99	−72	0.46
2016/2017	4.2	−19.0	9.2	−284	−245	0.65
2017/2018	9.7	−11.2	9.0	−107	−66	0.56
2018/2019*	7.1	−11.3	10.3	−125	−58	0.47

*Climatic characteristics recorded until March 1 2019

Table 2. Climatic characteristics based on the experimental stand measurements – measuring profile no. 5.

Winter period	$\theta_{s,max}$ (°C)	$\theta_{s,min}$ (°C)	θ_m (°C)	I_F (°C, day)	I_{FS} (°C, day)	D_F (m)
2016/2017	4.2	−19	9.2	−284	−283	0.63
2017/2018	9.7	−11.2	9.0	−107	−86	0.55
2018/2019*	7.1	−11.3	10.3	−125	−89	0.49

2.2 Analysis of the Experimental Monitoring Results at the Experimental Stand DRETM

It is possible to analyze the influence of the method of railway track routing (embankment, cut) since both measuring profiles were not built simultaneously (see Chapter 1). The analysis is based on the climatic characteristics for the winter periods 2016/2017 to 2018/2019 (see Tables 1 and 2), which represent a total of 3 winters.

Out of these 3 winter periods, the most adverse winter period, from the point of view of the obtained air frost index value I_F and the depth of the track substructure freezing D_F, was the winter 2016/2017. The value of the frost index I_F for the winter period 2016/2017 represented 60% of the design value of the frost index valid for the location of the experimental stand ($I_{F,d}$ = 470 °C, day location Žilina). This is the reason why this particular winter period was selected for the comparison of the influence of the track routing methods.

Figure 4 depicts the course of mean daily air temperatures θ_s and the method of evaluation of the air frost index value I_F at the Experimental Stand DRETM, valid for both measuring profiles. Air frost index I_F is determined by adding up mean values of the day air temperatures θ_s in the winter season. Mean daily air temperatures θ_s were determined based on the ambient air temperature as measured at 7 a.m., 2 p.m. and 9 p.m. of the Greenwich Mean Time at 2 m above the terrain.

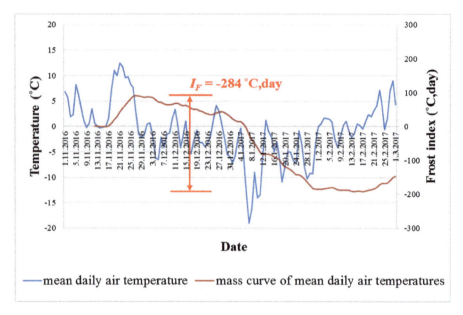

Fig. 4. Graphic representation of the course of mean daily temperatures and the air frost index for the winter period 2016/2017 (source: own research)

The course of mean daily temperatures on the ballast bed surface θ_{ss} and the surface frost index I_{FS} for the measuring profile no. 1 (embankment) is depicted in Fig. 5 and for the measuring profile no. 5 (cut) in Fig. 6.

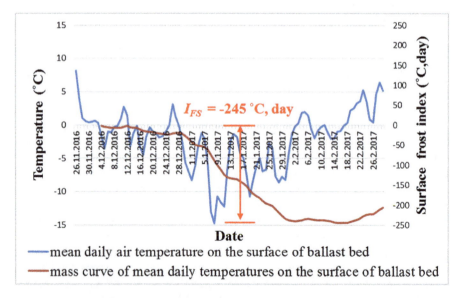

Fig. 5. Graphic representation of the course of mean daily temperatures and the frost index on the ballast bed surface for the winter period 2016/2017 (measuring profile no. 1 - embankment) (source: own research)

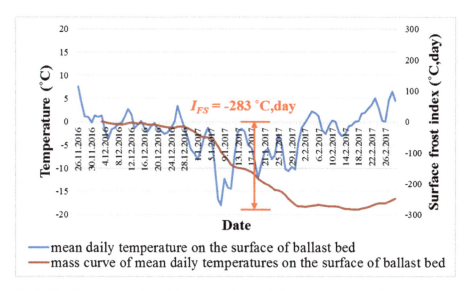

Fig. 6. Graphic representation of the course of mean daily temperatures and the frost index on the ballast bed surface for the winter period 2016/2017 (measuring profile no. 5 - cut) (source: own research)

The course of the depth of freezing of the railway track (track substructure) D_F recorded for the measuring profile no. 1 (embankment) is depicted in Fig. 7. and for the measuring profile no. 5 (cut) in Fig. 8.

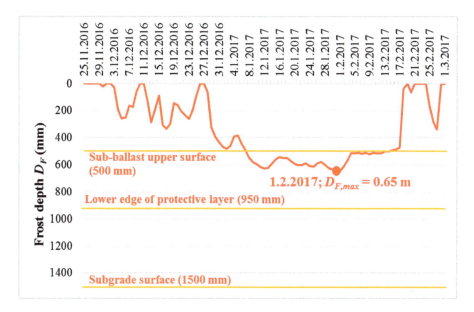

Fig. 7. Graphic representation of the course of the depth of freezing of the track substructure for the winter period 2016/2017 (measuring profile no. 1 – embankment) (source: own research)

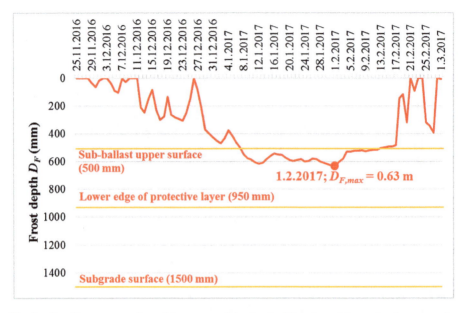

Fig. 8. Graphic representation of the course of the depth of freezing of the track substructure for the winter period 2016/2017 (measuring profile no. 5 – cut) (source: own research)

By comparing the values of the depths of the freezing of the track (track substructure) D_F for the measuring profile no. 1 (embankment) and for measuring profile no. 5 (cut), it is possible to conclude that the difference in the depths obtained for all the evaluated winter periods is up to 0.02 m. For the process of the track substructure dimensioning for the non-traffic load, this difference is a negligible one.

The differences in surface freezing indices determined on the ballast bed surface for the measuring profile no. 1 and the measuring profile no. 5, were larger, 20 to 40 °C, day (the effect of ground frosts for measuring profile no. 5 - cut).

It is interesting to note that these larger differences in achieved frost indices have virtually no effect on the reached depth of freezing of the track structure.

3 Conclusion

The monitoring of the influence of the track routing method (earthwork in the embankment or cut) at DRETM has been carried out since 2016 when the Experimental Stand DRETM was extended by the measuring profile no. 5 (cut). By comparing the climatic characteristics of the three recorded winter periods (2016/2017 to 2018/2019), it is possible to say that the shape of the railway body has an impact on the achieved value of the surface air frost index I_{FS} (due to the logical effect of larger ground frost in the cut) but it has a minimal effect on the final value of the track substructure freezing - up to 0.02 m.

Because in the track substructure dimensioning process for the non-traffic load [8] the designed thickness of the protective layer is rounded to 0.05 m upward, it is possible to neglect the influence of the track routing method.

The same result was achieved by the numerical modeling of the railway track structure using the *SoilVision* software, where the individual numerical models of the measuring profiles no. 1 and no. 5 were loaded with the design value of the air frost index [9].

Acknowledgements. The presented results are the results of solving the VEGA grant project 1/0275/16 *Optimization design of sleeper subgrade due to non-traffic load aspect*.

References

1. Newmann, G.P.: Heat and mass transfer in unsaturated soils during freezing. M.Sc. thesis. University of Saskatchewan, Canada (1995)
2. Pentland, J.S.: Use of a general partial differential equation solver for solution of heat and mass transfer problems in soils. University of Saskatchewan, Canada (2000)
3. Soliman, H., Kass, S., Fleury, N.: A simplified model to predict frost penetration for Manitoba soils. In: Annual Conference of the Transportation Association of Canada Toronto, Ontario (2008)
4. Tam, A.: Permafrost in Canada's Subarctic Region of Northern Ontario, University of Toronto (2009)
5. Fredlund, M.: SOILVISION, A Knowledge-Based Soils Database, User's Manual, Saskatoon, Saskatchewan, Canada (2011)
6. Thode, R.: SVHEAT, 2D/3D Geothermal Modeling Software, Tutorial Manual, Saskatoon, Saskatchewan, Canada (2012)
7. Ižvolt, L., Dobeš, P., Pultznerová, A.: Monitoring of moisture changes in the construction layers of the railway substructure body and its subgrade. Proc. Eng. **161**, 1049–1056 (2016). Elsevier B.V., Amsterdam, The Netherlands. ISSN 1877-7058
8. TNŽ 736312: The design of structural layers of subgrade structures (in Slovak). GR ŽSR. Slovakia (2005)
9. Ižvolt, L., Dobeš, P., Hodas, S.: Impact of the method of rail track routing on the thermal regime of subgrade structure – numerical modeling of non-traffic load. Int. J. Transp. Dev. Integr. **2**(3), 250–257 (2018). WIT Press, United Kingdom. ISSN 2058-8305

Advanced Methods for Data Collection and Analysis

Estimating Parameters of Demand for Trips by Public Bicycle System Using GPS Data

Vitalii Naumov and Krystian Banet(✉)

Transport Systems Department, Cracow University of Technology,
Krakow, Poland
{vnaumov,kbanet}@pk.edu.pl

Abstract. Bicycles become one of the main modes of public transport in modern cities. In many cases, they replace private cars and even public buses, especially in the cities with the developed bicycle infrastructure. However, designing and developing public transport systems should be implemented on the grounds of known parameters of demand for travels. The paper describes an approach to the travel demand estimation with the use of data obtained from GPS trackers. The article mainly focuses on data cleaning procedures and the estimations of the demand parameters on the base of the obtained dataset. The authors propose the software for reading raw records from GPX-files and cleaning the data. The case study of the bicycle share system in Kraków, Poland, is discussed in the paper: the results of the data cleansing and estimating demand parameters for recreational trips allocated from the obtained sample are shown.

Keywords: Public bicycle system · Demand parameters · Data cleansing · Recreational trips

1 Introduction

Urban planners and governments today are faced with the dilemma of providing high quality mobility services to a growing population, while at the same time minimizing energy consumption, reducing harmful environmental impacts and cultivating a lively and safe urban environment. As a means of meeting these challenges, the bicycle has resurfaced as a valuable transportation mode by Twaddle et al. (2014).

In spite of recommendations for conducting cycling studies, one of the trends mentioned by Zalewski (2009) is the creation of bicycle traffic systems, which constitute an integral part of the transport system of urban areas without making traffic forecasts. Pedestrian and bicycle traffic has been neglected for decades when preparing travel demand forecasts. Terner et al. (2006) point out that urban planners have started looking for ways to set priorities, and one of them is to anticipate demand or rely on current data. Until recently, achieving accurate estimates of the demand for bicycle traffic was very difficult due to the lack of data, but currently a sharp increase in the availability of data on bicycle travel can be observed.

Proulx and Pozdnukhov (2017) state that new comprehensive data sets as an addition to the classic methods include automated counters, crowdsource GPS tracks and data on the bicycles use. Kuzmyak and Dill (2012) point GPS path tracks as a promising technique for collecting data on bicycle traffic. Nair et al. (2013) mention that currently, one of the possibilities of acquiring a large amount of information on bicycle traffic is the analysis of data from urban bicycle rental systems. This data source is part of the definition of big data: according to the Cambridge Dictionary, big data are very large data sets that are created by people using the Internet and that can be stored, understood and used only by means of special tools and methods.

This papers aims to present the approach to the GPS data cleansing, proposed by the authors in order to obtain the purified sample for studies of demand for trips in bike-sharing systems, and to demonstrate the use of the obtained big volumes of data in procedures of the demand parameters estimations on the example of recreational trips completed in the bike-sharing system of Krakow.

The paper has the following structure: in the second part, we present a brief review of scientific publications in the field of bicycle demand studies with a focus on indicators influencing the demand parameters; the third section contains descriptions of the proposed data cleansing subroutines and the developed method for allocating the recreational trips in the purified sample; the forth part depicts results of the use of the developed methodology on example of the Wawelo bike-sharing system in Kraków; the last section proposes conclusions and directions for further research.

2 Literature Review

One of the pillars of the "smart" philosophy is modern and integrated transport; its essential element, which has become more and more important in recent years, not only in the largest metropolises but also in medium-sized cities and smallest local communities, are bike-sharing systems (Fall and Dąbrowski 2015). The potential of bicycle rental systems as a source of data on city dynamics and aggregate human behavior was indicated, among others, by Froehlich et al. (2009), who pay attention to the need to clean up the acquired data. GPS data acquired from the city bike system gives the possibility to analyze the parameters of city bikes travel demand, and also allow to examine factors that affect the selection of the route. Among the possible parameters, that should be tested, are the length of the journey, its duration, speed, and the lengthening factor (or the route elongation coefficient). Broach et al. (2012) proved on the example of Portland, that bicycle trips for recreation have an average length of 2.2 miles, while those for commuting to work or school 3.7 miles.

According to the data in the European Commission's publication by Dekoster and Schollaert (1999), in Europe, 30% of routes covered by a car are shorter than 3 km and 50% are shorter than 5 km. According to Pucher and Buehler (2008), for Europe, the large part of all trips is shorter than 2.5 km (44% in the Netherlands, 37% in Denmark, 41% in Germany). Hydén et al. (1999), in their studies of the cycling distance in Western European countries, showed that the average length of cycling trips is 2... 3 km depending on the source of data, except in the Netherlands and Denmark, where these journeys are longer due to pro-driving policies and flat terrain, but generally people do not use the bike as a means of transport for trip distances longer than 3 km.

Nair et al. (2013) conducted studies of cycling time, analyzing data from the bicycle rental system in Paris. They received information that 92% of the journeys are shorter than 30 min and 98% of the trips are shorter than 45 min. During the Warsaw Traffic Research 2015, Jacyna et al. (2016) obtained that 39.9% of bile trips are shorter than 15 min, 67.8% of journeys are shorter than 30 min, and trips lasting more than an hour constitute 2.9% of all bicycle trips in the city.

In addition to the distance and travel time, another parameter describing it is speed: instantaneous, average and communication. For more reliable data obtained from GPS devices, Broach et al. (2012) showed that the average communication speed of cyclists traveling for recreational purposes in Portland (USA) is 16.1 kmph, while for obligatory journeys equals 19.0 kmph. According to Nair et al. (2013), the average speed of cyclists in Western European countries is 10…25 kmph. Kopta and Rudnicki (1996) state that the communication speed of a bicycle can reach 10…15 kmph.

The other parameter, frequently analyzed in the literature, is the route elongation coefficient. According to the recommendations of the Dutch government, for the main cycling routes, it should be at the level of 1.2, for collecting routes – 1.3, and for the "last mile" routes – 1.4. According to the report of Centre for Research and Contract Standardization in Civil and Traffic Engineering, studies of route selection schedules conducted in the Netherlands showed that 50% of cyclists choose routes that are longer than 6% of the shortest route. An example of research conducted in Poland regarding the elongation factor in bicycle trips was the research by Rakower et al. (2011). They investigated that in Poznań the elongation rate is 1.22, which is similar to the Dutch recommendations.

Among the factors that affect cycling demand, Sener et al. (2009) list 6 groups of indicators that can be taken into account when considering the route selection by cyclists. The first group is represented by characteristics describing the cyclist, such as age, sex, occupation, experience in cycling. The second group is formed by characteristics related to street parking, i.e. type of parking, rotation indicator, parking level indicator or the length of the parking area. Another group are factors related to cycling facilities, among which Sener et al. (2009) mention the continuity and type of the route, i.e. whether it is a separate road for bikes or a wide outside lane that allows movement of two cyclists and a car next to each other or a cyclist and a lorry. The fourth group is the physical characteristics of the road, such as the road class, the number of "stop" signs, traffic lights and intersections. The fifth group consists of functional road attributes: traffic volume and speed limit, and the last group are operational characteristics of the road, e.g. travel time. The abovementioned factors were studied in literature by various researchers (see Table 1).

The analysis of the above studies shows that the most often mentioned parameter affecting the selection of the bicycle route is the type of route. In none of the studies did the influence of cultural and natural attractions on the route selection appear. In this research, on the base of data from GPS-trackers, we intend to allocate recreational journeys, implemented within the bike-sharing system, and to show that parameters of such trips differ significantly from the according parameters of the whole dataset.

Table 1. Parameters that influence the parameters of bicycle trips.

Indicator	Authors
Age	Antonakos (1994), Hopkinson and Wardman (1996), Aultman-Hall (1996), Stinson and Bhat (2003), Hunt and Abraham (2007), Tilahun et al. (2007)
Sex	Antonakos (1994), Hopkinson and Wardman (1996), Aultman-Hall (1996), Stinson and Bhat (2003), Tilahun et al. (2007)
Income	Stinson and Bhat (2003), Tilahun et al. (2007)
Access to the car	Antonakos (1994), Hopkinson and Wardman (1996)
Access to the bicycle	Antonakos (1994), Hopkinson and Wardman (1996)
Type of bicycle route	Lott et al. (1978), Bovy and Bradley (1985), Axhausen and Smith (1986), Antonakos (1994), Davis (1995), Guttenplan and Patten (1995), Aultman-Hall (1996), Landis et al. (1997), Stinson and Bhat (2003), Hunt and Abraham (2007), Tilahun et al. (2007), Broach et al. (2012)
Continuity of the bicycle route	Antonakos (1994), Stinson and Bhat (2003)
Surface of the bicycle route	Bovy and Bradley (1985), Axhausen and Smith (1986), Antonakos (1994), Davis (1995), Landis et al. (1997)

3 Methodology

3.1 Data Cleansing Subroutines

The data obtained from GPS-trackers is being stored in files with GPX-format – an XML schema designed as a common GPS data format. In order to read the data from the GPX-files, we have developed the code in Python, which allows presenting the data as a set of the trip parameters – the identification number of the trip, the number of track segments, the travel duration, the idle time, the total trip distance, and the average travel speed. Presenting the bicycle trips data in such format have allowed us to obtain samples of random variables characterizing numeric parameters of demand.

The proposed data cleansing methodology contains three main stages:

- stage A: eliminating the damaged data: the raw data set usually contains records with damaged data caused by interruptions of the signal of GPS-trackers, such signal interferences yield in unrealistic timing of bicycle travels;
- stage B: excluding the trips with parameters equal to zero: some tracks contain zero values of the trip duration or the trip distance; usually, such data records represent the situations when the bicycle was locked off the rack at the station but wasn't used;
- stage C: excluding the trips which were not completed: some tracks are characterized by non-zero travel duration but extremely small (less than 50 m) trip distance; these records are caused by the cases when the bike was locked off the rack and then left near the station or locked in the rack after some time without being used for travel.

Performing the listed steps on the raw track records allows purifying the data: the obtained set will contain only accomplished trips representing the real demand for travels by bicycle.

The developed code for data reading and the corresponding cleansing and parameters' estimating procedures could be downloaded from the git-repository available at https://github.com/naumovvs/city-bikes-analysis.git.

3.2 Allocating the Recreational Trips on the Base of GPS Data

Using the purified data, different types of trips within the bike-sharing system can be allocated for the given purpose of journey, such as traveling to or from work, shopping trips, visiting friends or relatives, recreational activities, etc. The division of trips according to categories of purposes is one of essential stages in travel demand modelling procedures, so such segmentation implemented automatically, by applying data analysis procedures, gives a valuable information. In the frame of this study, we develop the methodology for allocating the trips within the bike-sharing system for the recreational purposes.

By using the cleaned dataset, it's decided to examine the impact of the attractiveness of route surroundings on the travel parameters implemented within the city bike system in Kraków. These route surroundings influence the route shape for recreational trips: the route trace is chosen by travelers in a way to locate it near one or more of cultural or natural attractions. The attractiveness of the route's surroundings can be expressed through the attractions described by the relevant parameters, such as:

- the grade given by tourists,
- location of the place,
- the number of other attractions in the neighborhood,
- radius of visual impact of the attraction,
- a type of the attraction (cultural or natural), etc.

The parameters of the bicycle route surroundings can be obtained from crowd-sourcing information systems, among which, nowadays, the Google Maps is the most popular (that means – containing the biggest amount of publicly available data).

On the base of user recommendations in the Google Maps, the clusters of cultural and natural attractions were defined for the Śródmieście district of the Kraków city. The map of clusters along with marked zones of the visual impact of attractions is presented on Fig. 1.

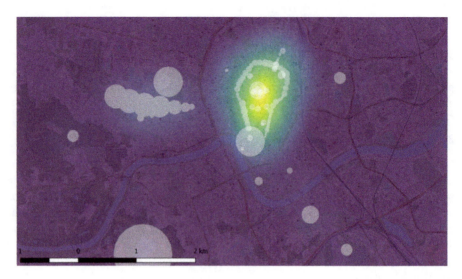

Fig. 1. Clusters of analyzed cultural and natural attractions within the Kraków city center

In total, more than 40 attractions were selected in the Śródmieście district. For the purposes of this study, five cultural attractions were used to investigate the parameters of recreational trips (characteristics of the selected attractions are presented in Table 2).

Table 2. Tourist cultural attractions in Kraków selected for preliminary analysis.

Attractions	Average grade (checked on 26.02.2019)	Number of opinions (checked on 26.02.2019)	Longitude coordinate	Latitude coordinate
Barbican	4.6	1,306	19.941631800681	50.0655252179242
The Corpus Christi Basilica	4.7	503	19.9449736250858	50.0498022291887
Saint Mary's Church	4.8	3,945	19.9394150364056	50.061695256167
The Collegiate Church of St. Florian	4.6	227	19.943279371554	50.0675565550794
St. Stanislaus Church	4.7	478	19.9376004584569	50.0482450615404

The initial and crucial phase of investigating the impact of the route surroundings on the trip type is to check whether a statistically significant part of the journeys within the city bike system can be classified as recreational trips. The allocation of recreational trips on the base of analysis of the route's surroundings has been divided into three main stages:

- stage I: separation of journeys taking place near the cultural and natural attractions (for each of the analyzed attractions) – within the radius of the visual impact of the cultural or natural attraction,
- stage II: selections from the trips obtained in stage I of all the trips with a route elongation coefficient value greater than 2.0,
- stage III: comparison of travel demand parameters extracted in stage II with parameters from the entire sample of the purified data.

The second phase was the key stage; its aim was to choose such trips, for which the route trace is close to the cultural or natural attraction (within the zone of the attraction visual impact), but the route chosen by the traveler is not the shortest; that allows us to hypothesize that the cultural and natural attraction is a factor influencing the choice of route and the trip can be classified as recreational. In addition to the route elongation coefficient, other explanatory variables can be used, such as the route length and the average trip speed.

Implementation of the developed data analysis procedures related to allocation of recreational journeys can be also found in the repository at https://github.com/naumovvs/city-bikes-analysis.git.

4 Results and Discussion

4.1 Data Cleansing Results

The proposed methodology of the data cleansing and estimations of the demand parameters were used on the example of Wavelo bike sharing system in Kraków, Poland. In total, 34,969 tracks, obtained from GPS-trackers in the period between 31 May and 7 June 2017, were analyzed. After the stage A, 5,946 damaged records were dropped out of consideration. At the stage B, 675 tracks were excluded from the sample – 40 records with zero trip duration and 635 records with zero trip distance, these are the cases when bikes were chosen at the stations but were not shifted from the rack. Finally, after the stage C, 421 records representing the trips with the total travel distance less than 50 m were excluded. As the result, the sample of 27,927 records containing data on the actually performed travels was obtained.

The performed procedure of the sample purification is illustrated on Figs. 2, 3, 4 and 5 on the example of the average speed distribution. As it could be seen, the purer the data in the analyzed dataset, the average speed distribution become closer to Gaussian shape.

Figure 2 shows a histogram of the average trip speed for the raw data. In the presented graph we can point out that a significant part of trips is characterized by extremely low speed (equal or very close to 0). Additionally, there's an abnormal number of journeys with the average speed up to 7…8 kmph that results in distortion of the histogram shape and its deviation from the normal distribution.

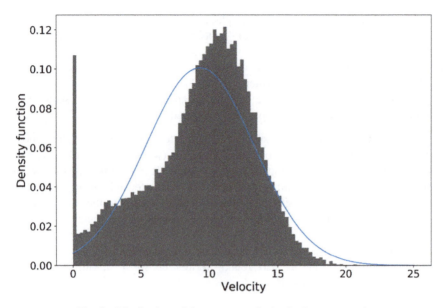

Fig. 2. Distribution of the average velocity in the raw sample

After the stage A, the shape of the average speed distribution becomes much closer to the Gaussian, but the histogram still contains values close to zero (see Fig. 3), shifting the fitted theoretical normal curve to the left from the empirical graph.

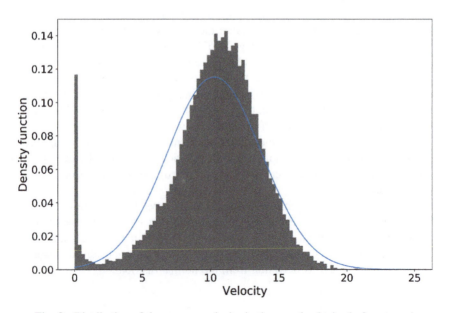

Fig. 3. Distribution of the average velocity in the sample obtained after stage A

Removal of journeys with zero duration and zero travel distance yields the histogram presented in Fig. 4. It amends the histogram shape, but the theoretical curve remains shifted a bit to the left.

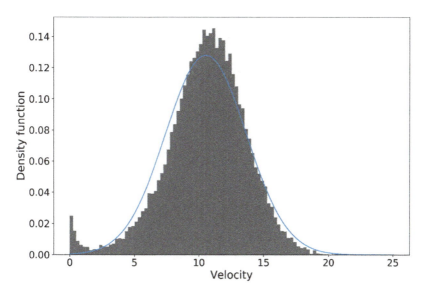

Fig. 4. Distribution of the average velocity in the sample obtained after stage B

The last stage of the data cleansing procedure is eliminating of the uncompleted trips from the sample: as a result, for the obtained dataset, the average speed obeys the normal distribution (see Fig. 5).

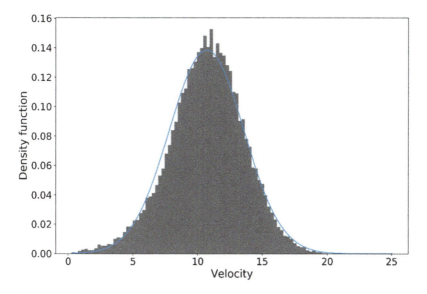

Fig. 5. Distribution of the average velocity in the raw sample obtained after last stage

For the cleansed sample, the distribution of the trip distance and the travel duration was substantiated: the hypothesis on the gamma-distribution was not rejected for both parameters by the Pearson's chi-squared test (the 95% confidence level was used while testing the hypotheses on the parameter's distribution). The mean value of the trip distance equals to 3.82 km for the raw data and 3.89 km for the purified sample, while the mean travel duration in the initial sample equals 0.48 against 0.39 h for the cleansed data.

For the trips in the cleaned sample, values of the route elongation coefficient were also examined. The obtained distribution of the coefficient is presented in Table 3.

Table 3. Distribution of the route elongation coefficient in the cleaned sample.

Value of the route elongation coefficient	Percentage of trips which route elongation is less than the given value [%]
1.2	25.7
1.3	44.0
1.5	69.7
2.0	84.9
3.0	89.4

As can be concluded from the provided distribution, the median of the route elongation coefficient is located in the range between 1.3 and 1.5. Taking into account the existing road network, in this research, we assume that the journey in the Wavelo bike sharing system is implemented by the elongated trace (with the route length much bigger than the shortest one) if the coefficient value is greater than 2.0.

4.2 Data Cleansing Results

Analysis of data from the Wavelo system in accordance with the proposed methodology has shown that a statistically significant part of travels within the city bike system is performed for recreational purposes, assuming that such journeys have the route prolongation coefficient greater than 2.0 (see Table 4).

Table 4. Parameters of recreational trips running nearby the analyzed attractions.

Tourist attraction	Number of bicycle trips running nearby the attraction	Percentage of the trips with the route elongation coefficient greater than 2.0 [%]	The average speed with for trips with the route elongation coefficient greater than 2.0 [kmph]
Barbican	1502	17.5	8.50
The Corpus Christi Basilica	361	56.7	7.86
Saint Mary's Church	829	18.3	7.67
The Collegiate Church of St. Florian	632	16.6	8.57
St. Stanislaus Church	350	31.4	8.76

Thus, initial analysis has shown that the characteristics of the route surroundings affect the travel parameters within the city bike system. In this way, recreational trips within a bike-sharing system can be distinguished by analyzing data from GPS trackers. The distinction of the allocated set of journeys can be shown on the basis of the average speed, which for the entire sample set equals 10.70 kmph, and in the case of journeys qualified as recreational for the analyzed five cultural and natural attractions ranges from 7.67 to 8.76 kmph.

5 Conclusions

The article allows for partial completion of the research gap, which concerns the estimations of demand for bike-sharing systems. The proposed approach to the estimations of demand parameters uses modern methods of data collection and its analysis. As a result, a large sample of good quality data can be obtained; that allows researchers not only to evaluate characteristics of demand but also to allocate segments of demand with the given features (e.g., the trip purposes can be distinguished from the dataset).

In this paper, as an example, we managed to allocate the segment of recreational journeys by studying characteristics of the route surroundings that affect the travel parameters within the city bike system. As a result, we can state that parameters of the analyzed cultural and natural attractions (including location and assessment by tourists) have an impact on the average speed of travel by city bikes, which differed significantly from the average speed in the entire research sample.

As for directions of future research, the use of the proposed methodology for the bigger amounts of data should be mentioned. Developing the subroutines for allocating other trip purposes than recreational journeys must be considered to provide complete information for demand modeling procedures.

References

Antonakos, C.L.: Environmental and travel preferences of cyclists. Transp. Res. Rec: **1438**, 25–33 (1994)

Aultman-Hall, L.: Commuter Bicycle Route Choice: Analysis of Major Determinants and Safety Implications, p. 224. McMaster University, Ontario (1996)

Axhausen, K.W., Smith, R.L.: Bicyclist link evaluation: a stated-preference approach. Transp. Res. Rec. J. Transp. Res. Board **1085**, 7–15 (1986)

Bovy, H.L., Bradley, M.A.: Route choice analyzed with stated-preference approaches. Transp. Res. Rec. **1037**, 11–20 (1985)

Broach, J., Dill, J., Gliebe, J.: Where do cyclists ride? A route choice model developed with revealed preference GPS data. Transp. Res. Part A Policy Pract. **46**(10), 1730–1740 (2012)

Davis, W.J.: Bicycle test route evaluation for urban road conditions. In: Transportation Congress: Civil Engineers – Key to the World's Infrastructure, vol. 2, pp. 1063–1076 (1995)

Dekoster, J., Schollaert, U.: Cycling: The Way Ahead for Towns and Cities, p. 63. Office for Official Publications of the European Communities, Luxembourg (1999)

Fall, M., Dąbrowski, M.: Jak rowery miejskie tworzą "smart cities". In: Biała Księga Mobilności, pp. 118–121 (2015)

Froehlich, J., Neumann, J., Oliver, N.: Sensing and predicting the pulse of the city through shared bicycling. IJCAI Int. J. Conf. Artif. Intell. **3**, 1420–1426 (2009)

Guttenplan, M., Patten, R.: Off-road but on track: using bicycle and pedestrian trails for transportation. TR News **178**(3), 7–11 (1995)

Hopkinson, P., Wardman, M.: Evaluating the demand for new cycle facilities. Transp. Policy **3**(4), 241–249 (1996)

Hunt, J.D., Abraham, J.E.: Influences on bicycle use. Transportation **34**(4), 453–470 (2007)

Hydén, C., Nilsson, A., Risser, R.: How to enhance WALking and CYcliNG instead of shorter car trips and to make these modes safer. Lund University Faculty of Engineering, Technology and Society, Transport and Roads, Lund, p. 68 (1999)

Jacyna, M., Wasiak, M., Gołębiowski, P.: Model ruchu rowerowego dla Warszawy według Warszawskiego Badania Ruchu 2015. Transport Miejski i Regionalny **10**, 5–11 (2016)

Kopta, T., Rudnicki, A.: Planistyczno-realizacyjne aspekty komunikacji rowerowej. Transport Miejski, 5–6 (1996)

Kuzmyak, J.R., Dill, J.: Walking and bicycling in the united states: the who, what, where, and why. TR News **280**, 4–15 (2012)

Landis, B.W., Vattikutti, V.R., Brannick, M.: Real-time human perceptions: towards a bicycle level of service. Transp. Res. Rec. **1578**, 119–126 (1997)

Lott, D.Y., Tardiff, T., Lott, D.F.: Evaluation by experienced riders of a new bicycle lane in an established bikeway system. Transp. Res. Rec. **683**, 40–46 (1978)

Nair, R., Miller-Hooks, E., Hampshire, R.C., Bušić, A.: Large-scale vehicle sharing systems: analysis of Vélib. Int. J. Sustain. Transp. **7**(1), 85–106 (2013)

Proulx, F.R., Pozdnukhov, A.: Bicycle traffic volume estimation using geographically weighteddata fusion. Preprint (2017). http://faculty.ce.berkeley.edu/pozdnukhov/papers/Direct_Demand_Fusion_Cycling.pdf

Pucher, J., Buehler, R.: Cycling for everyone: lessons from Europe. Transp. Res. Rec. J. Transp. Res. Board **2074**(12), 58–65 (2008)

Rakower, R., Łabędzki, J., Gadziński, J.: Konkurencyjność ruchu rowerowego w przestrzeni miejskiej. Transport Miejski i Regionalny **2**, 31–38 (2011)

Sener, I.N., Eluru, N., Bhat, C.R.: An analysis of bicycle route choice preferences in Texas, US. Transportation **36**(5), 511–539 (2009)

Sign up for the bike: design manual for a cycle-friendly infrastructure. Centre for Research and Contract Standardization in Civil and Traffic Engineering. The Centre, Ede, p. 325 (1993)

Stinson, M.A., Bhat, C.R.: An analysis of commuter bicyclist route choice using a stated preference survey. Transp. Res. Rec. J. Transp. Res. Board **1828**(1), 107–115 (2003)

Tilahun, N.Y., Levinson, D.M., Krizek, K.J.: Trails, lanes, or traffic: valuing bicycle facilities with an adaptive stated preference survey. Transp. Res. Part A Policy Pract. **41**(4), 287–301 (2007)

Turner, S., Sandt, L., Toole, J., Benz, R., Patten, R.: FHWA University Course on Bicycle and Pedestrian Transportation: Student Workbook. US Department of Transportation, p. 453 (2006)

Twaddle, H., Schendzielorz, T., Fakler, O.: Bicycles in urban areas. Transp. Res. Rec. J. Transp. Res. Board **2434**(1), 140–146 (2014)

Zalewski, A.: Modele ruchu rowerowego w miastach i aglomeracjach. In: Zeszyty Naukowo-Techniczne Stowarzyszenia Inżynierów i Techników Komunikacji w Krakowie. Seria: Materiały Konferencyjne: Ogólnopolska Konferencja Naukowo-Techniczna Modelowanie podróży i prognozowanie ruchu. pp. 263–275 (2009)

Support for Pro-ecological Solutions in Smart Cities with the Use of Travel Databases – a Case Study Based on a Bike-Sharing System in Budapest

Katarzyna Turoń[1(✉)], Grzegorz Sierpiński[1], and János Tóth[2]

[1] Faculty of Transport, Silesian University of Technology, Katowice, Poland
{katarzyna.turon,grzegorz.sierpinski}@polsl.pl
[2] Department of Transport Technology and Economics,
Budapest University of Technology and Economics, Budapest, Hungary
toth.janos@mail.bme.hu

Abstract. The work was dedicated to the topic of "bike-sharing systems", or short-term bicycle rentals, available in modern and smart cities. The aim of the article is to present how travel databases can facilitate management and development of bike-sharing solutions. The article is based on a case study of a bike-sharing system in Budapest. The article also includes a statistical analysis based on data collected in 2017 and 2018. Research results on travelling in specific years enable, among others, to predict seasonal changes, as well as determine the attractiveness of individual docking stations. The article contains selected research results in this area.

Keywords: Bike-sharing · Bike-sharing systems · Travel databases · Smart city

1 Introduction

Transport is one of major concerns in contemporary cities. Therefore, local governments strive to improve the efficiency of travelling within the city. Considering guidelines for the future development of transport systems [1–3], improvements should be made regarding such issues as [4–10]:

- mitigating negative impact of transport on the environment,
- changing modal split,
- reducing energy consumption,
- improving land use, and
- ensuring traffic safety.

The smart city approach requires data to be collected and tools developed to support the decision-making process. For this reason, cities monitor and measure traffic and transport infrastructure (inter alia [11–15]). The next step requires methods that enable to use data collected in decision-making (e.g. [16–21]). Measurements are rarely automatic and data acquired are insufficient or poorly reflect actual needs of citizens (e.g. travel needs).

The article draws attention to the possibility of using modern database systems operated by urban bike rentals. For example, the bike rental system in Budapest, Hungary, provides basic information which is automatically stored in the database.

2 Bike-Sharing as a Form of Environmentally-Friendly Travelling

There are many different solutions intended to improve the current status of urban transport while eliminating the negative impact of traffic on the environment and society. However, after travelling by foot, cycling is the most non-polluting and healthy form of transport [22]. In the past, anyone who wanted to travel by bike needed to have one. This often meant that the person had to incur high cost of purchasing, maintenance and storage. An alternative to the above was to rent a bike. However, the first bike rentals offered long-term hire only. A customer who wished to make only a few minute ride had to pay an all-day fee for the use of a bike. Additionally, hiring a bike required a lengthy and complicated procedure. Customers had to visit the bike rental and fill out a number of forms.

The development of the shared economy created a possibility of possessing "something" without the need to buy it. In line with the trend, several operators started offering short-term rental of means of transport on the market. The trend rapidly evolved in the world leading to "bike-sharing", i.e. short-term bike rental. Now, new technologies are used to support rental of means of transport, with more than 1600 bike-sharing systems operating in 23 countries all over the world [23]. It is possible to hire approximately 15,284,850 classic and electric bicycles from over 330 operators [24].

Although bike-sharing systems are still considered innovative in urban mobility, they are becoming increasingly popular in the world. In fact, the short-term bike rental concept is nothing new. The first systems in the world started in 1965 in Amsterdam with a pioneering system called the White Bicycle Plan [25]. In consecutive years, a growing number of short-term bike rentals developed [26, 27].

The development of technology, opportunities created by mobile phones, and then on-line services led to the emergence of further generations of bicycle rental systems. These systems have been improved in terms of conditions to rent, type of a system that generates information about the possibility to rent a bike at docking stations or dock-less systems, areas of operation, modern bike fleet, including electric bicycles, bicycles for children or rickshaws, methods of payment, price, etc. [28]. The literature refers to 5 main generations of bike-sharing systems: [29–32]:

(1) First generation - the first commonly available bike-sharing system launched in Amsterdam in 1965. Pioneering bike-sharing system. Bicycles were made available for free, but due to the large number of thefts the system ceased to operate shortly after it had been launched.
(2) The second generation - a system launched in Copenhagen, Denmark, in 1991. It was the first system that provided short-term bike rental against a deposit.

(3) The third generation includes systems that operated globally in 1996–2014. Then, due to the rapid development of telecommunication technology and magnetic cards, a new system developed based on magnetic cards that could be topped up with pulses to be later used during trips. Moreover, the system enabled to open and close a bike station automatically. It was a very modern form of bike-sharing in comparison with previous ones.
(4) The fourth generation includes bike-sharing systems referred to as smart. By using the possibilities offered by smartphones and ITS technologies, bike-sharing systems started to use mobile applications for booking a bike from the level of a mobile device, as well as checking the availability of bicycles in a docking station or real-time monitoring of the location of a bike.
(5) The fifth generation is currently the most advanced technology-wise bike-sharing system on the market. Instead of a docking station, the use of the big data management enables to leave bicycles at any location designated by the operator. Bicycles offered in such systems are referred to as "dock-less" bikes.

With the increase in popularity of bike-sharing systems, cities and metropolitan areas have implemented new generations of services. Major interest of the public in travelling by bicycles triggered infrastructural changes, e.g. new cycling tracks, more attention given to the availability and status of pavements, and the designation of urban areas inaccessible for vehicle traffic [33–35].

Benefits of cycling have been for years highly appreciated in such countries as Denmark, the Netherlands and China. These are locations with an advanced cycling culture [36]. Now, in line with the sustainable development policy, the European Union has focused on the promotion of cycling [37] and highlighting the bicycle as the most sustainable mode of urban transport that can be used for short and long trips [38].

Considering the popularity of bike-sharing systems, it is important to monitor their use to ensure the development of the service as an alternative mode of transport in cities. Therefore, the authors have analysed bike-sharing in Budapest, one of major European capital cities.

3 Case Study

3.1 Description of BuBi MOL, a Bike-Sharing System Operating in Budapest

The practical part of the paper includes analyses based on the case study of BuBi MOL, a system operating in Budapest. The name of the system is an acronym derived from the combination of two Hungarian words BUdapest and BIcikli (which mean "bicycle" in Hungarian). The system has been operating in the Hungarian capital since 8 September 2014. It is managed by the BKK Center for Budapest Transport.

Additionally, the operator provides short-term rental of MOL Limo cars. At the beginning, the system consisted of 76 docking stations and 1100 bikes. The operation area was 15 km^2 in the city of Budapest. The operator invested EUR3.5 million in the system. As much as 85% of the amount was provided by the EU.

After BuBi MOL, the Donkey Republic developed as the second bike rental system operating in Budapest. Basic information about the two systems is shown in Table 1.

Table 1. Bike-sharing systems in Budapest (source: own research).

Feature	BuBi MOL	Donkey republic
Type of bicycles	Traditional bicycles	Traditional bicycles
Type of system	Station-based and station-free	Station-based and station-free
Colour and visibility	bright-green bicycles	bright-orange bicycles
Availability for user	Mobile App, MOL Bubi Pass, Bubi Ticket	Donkey Republic App
Service Packages	Yes	No

The system operates characteristic bright-green bicycles and the colour is an important traffic safety factor. A BuBi MOL docking station together with bicycles is shown in Fig. 1.

Fig. 1. Example of bike-sharing station in Budapest (source: own research)

From the beginning of its operation, BuBi MOL invested in the development of the bicycle fleet. In the first year of its operation, the system offered 1100 bikes. Today, the number of bikes has grown to 1846. Detailed information about the number of bicycles in consecutive years is shown in Fig. 2.

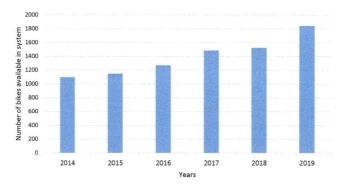

Fig. 2. Number of bikes available in the bike-sharing system in 2014–2019 (source: own research based on external data source)

Apart from bikes, the operator has invested in the number of docking stations as well. From the beginning of its operation, 67 new stations have been added to the system to make the total of 143 docking stations. Detailed information about the number of docking stations in consecutive years is shown in Fig. 3.

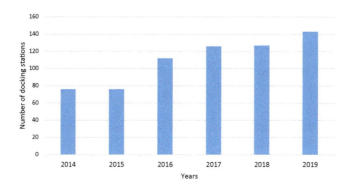

Fig. 3. Number of docking stations in the bike-sharing system in 2014–2019 (source: own research based on external data source)

BuBi MOL offers differentiated price schemes for its customers. To become a user, a person needs to buy a ticket or a pass for a specific period. The following schemes are available:

- 24 h,
- 72 h,
- Weekly,
- Quarterly,
- Semi-annual,
- Annual.

Ticket prices are presented in Table 2.

Table 2. Ticket prices by BuBi MOL tariffs (source: own research based on MOL information desk).

Tariff	Price in HUF	Price in EUR*
24 h	500	1.55
72 h	1,000	3.10
Weekly	2,000	6.20
Quarterly	5,000 (promo price); 7,800 (regular price)	15.50 24.18
Semi-annual	8,000 (promo price); 12,500 (regular price)	24.80 38.75
Annual	12,000 (promo price); 18,900 (regular price)	37.20 58.59

*Currency 1HUF = .0031 EUR (24th July 2019)

Regardless the tariff, the first 30 min of each ride is free, provided the user has an active ticket. After 30 min, the user is charged for every minute, and the fee gradually increases with time.

3.2 Variability in Bike Rental in Time

The structure of the bike rental data base specifies inter alia:

- start of bicycle rental,
- end of bicycle rental,
- name of the docking station from which the bicycle has been rented,
- name of the docking station to which the bicycle has been returned,
- number of bicycle rented,
- customer number,
- customer tariff etc.

Table 3 below contains summary results of an analysis that covers two years (2017 and 2018). According to the table, the number of bike rentals decreased by about 60 thousand. The poorest seasonal interest in renting a bike was in winter, whereas the highest number of rentals was noted in other months: in 2017 in July and in 2018 in May (Fig. 4). At the same time, it should be noted that data were recorded at very high accuracy levels. The number of incomplete records accounted for 1% in 2017. In 2018, the accuracy of data further increased. Therefore, the data collected fully reflect the behaviour of people renting city bikes, and such data can be used as a basis for decision-making regarding e.g. changes in the location docking stations.

Table 3. Leading particulars of city bike rental systems in 2017 and 2018 (source: own research).

Particulars	2017	2018
Total number of rentals	531,767	470,016
Average number of rentals per month	44,314	39,168
Minimum number of rentals per month	9,839	9,790
Maximum number of rentals per month	76,475	70,028
Average bike use [in minutes]	16.47	16.41
Percentage of incomplete records in database	1.019	0.450

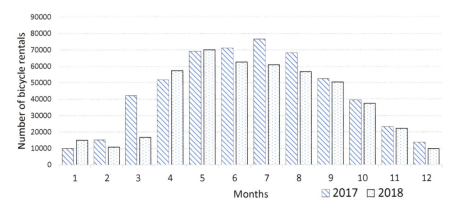

Fig. 4. Distribution of bike rentals in particular months of 2017 and 2018 (source: own research)

The average bike rental time (Table 3) is approximately 16 min. The distribution of average rental periods in particular months shows a slight decrease in winter months (Fig. 5). A large proportion of the total are trips shorter than 30 min, i.e. free trips. This means that the idea of using a city bicycle as an alternative to other modes on shorter distances has proved efficient.

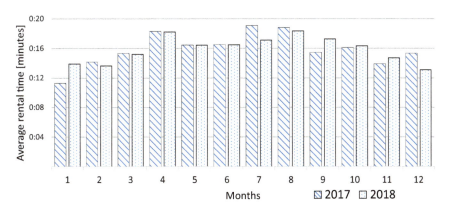

Fig. 5. Distribution of average rental periods by particular months in 2017 and 2018 (source: own research)

The analysis of the statistics with breakdown by docking stations shows a noticeable change in the value of the kurtosis. In 2018, the data concentration increased (Fig. 6). The measure of asymmetry has a positive value (Table 4), which means a right skewness, or a distribution in which the graph's peak leans to the left and its tail is extended to the right. Thus, for the skewness to be positive, the majority of data have values below the average, which is compensated by large positive values adopted by a few elements. Consequently, 2018 figures significantly improved.

Table 4. Distribution of docking stations and number of rentals in particular months in 2017 and 2018 (source: own research).

Statistics	2017	2018
Average number of rentals per docking station	4,211	3,743
Maximum number of rentals per docking station	12,309	12,237
Coefficient of variation	0.68	0.64
Skewness	0.96	1.21
Kurtosis	0.56	1.53

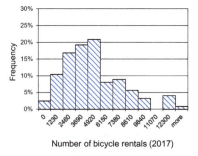
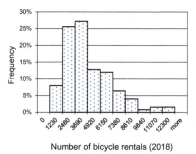

Fig. 6. Distribution of docking stations and number of rentals in 2017 and 2018 (source: own research)

3.3 Variability in Bike Rental in Space

The identification of rental and return stations allows to analyse the spatial distribution, as well as the attractiveness of docking stations and adjacent areas. The use of GIS tools helped to determine differences in the total number of rentals and returns in individual stations in the two years covered by the analysis. Results are presented in Figs. 7 and 8. The larger the circle, the larger number of rentals/returns. In general, a reduced interest in docking stations situated furthest to the north can be noted, whereas a major interest in bicycle rental can be observed in the central and eastern part of the rental area.

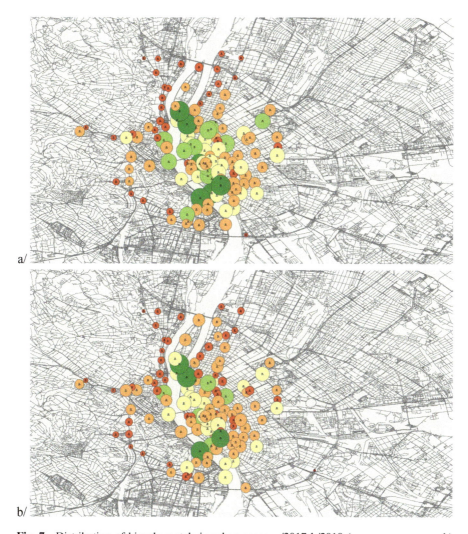

Fig. 7. Distribution of bicycle rentals in urban space: a/2017 b/2018 (source: own research)

By comparing the number of rentals in Fig. 7, we can see a slight increase in the interest in the north-eastern part of the city centre (where new docking station has been established) and on the island. Other stations recorded a slight decrease in the number of rentals. A similar situation has been noted when analysing the number of returns in individual docking stations (Fig. 8).

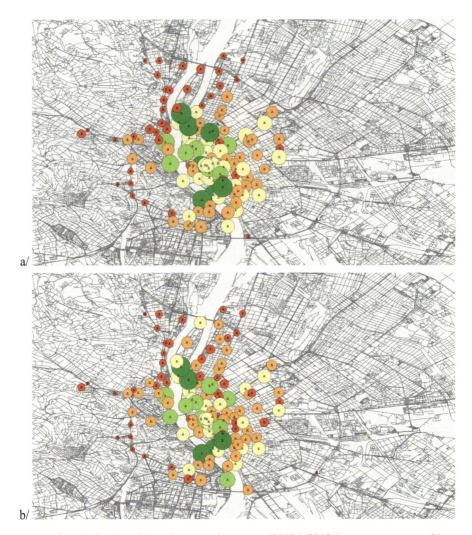

Fig. 8. Distribution of bicycle returns in space: a/2017 b/2018 (source: own research)

4 Summary

The access to traffic databases is necessary to examine the trips made. The goal of the article is to draw attention to the possibility of using steadily growing volumes of data on the sparing services market. Such data can be used not only by companies for commercial purposes, but also by cities to improve operation, planning of the transport system, and the creation of real alternatives to less environmentally friendly modes of transport.

As a part of further research, we plan to identify the spatial distribution of traffic and intend to determine the relationship between the location of a docking station and the attractiveness of a specific area.

Acknowledgments. This article is a part of the project entitled "The functioning of bike-sharing systems in Budapest and Warsaw" financed by the Visegard Fund under the Intra-Visegrad Scholarship in the academic year of 2018/2019. The authors should thank Mr Péter Dalos, the Bubi Bike Sharing Product Manager, from BKK Center for Budapest Transport for making data available for the purpose of analysis.

Visegrad Fund

References

1. White Paper: Roadmap to a Single European Transport Area – Towards a competitive and resource efficient transport system. COM, vol. 144 (2011)
2. Communication From The Commission to the European Parliament, The Council, The European Economic and Social Committee and The Committee of the Regions: Clean Power for Transport: A European alternative fuels strategy, COM, vol. 17 (2013)
3. White Paper on the Future of Europe, Reflections and scenarios for the EU27 by 2025, COM, vol. 2025 (2017)
4. Jacyna, M., Zak, J., Jacyna-Gołda, I., Merkisz, J., Merkisz-Guranowska, A., Pielucha, J.: Selected aspects of the model of proecological transport system. J. KONES, Powertrain Transp. **20**, 193–202 (2013)
5. Kauf, S.: City logistics - a strategic element of sustainable urban development. Transp. Res. Procedia **16**, 158–164 (2016)
6. Santos, G.: Road transport and CO_2 emissions: what are the challenges? Transp. Policy **59**, 71–74 (2017)
7. Celiński, I.: GT Planner used as a tool for sustainable development of transport infrastructure. In: Suchanek, M. (ed.) New Research Trends in Transport Sustainability and Innovation: TranSopot. In: 2017 Conference Springer Proceedings in Business and Economics, pp. 15–27 (2018)
8. Galińska, B.: Multiple criteria evaluation of global transportation systems - analysis of case study. In: Sierpiński, G. (ed.) Advanced Solutions of Transport Systems for Growing Mobility. Advances in Intelligent Systems and Computing, vol. 631, pp. 155–171 (2018)
9. Staniek, M., Czech, P.: Self-correcting neural network in road pavement diagnostics. Automation in Construction, vol. 96, pp. 75–87. Elsevier, Amsterdam (2018)
10. Galińska, B.: Intelligent decision making in transport. evaluation of transportation modes (types of vehicles) based on multiple criteria methodology. In: Sierpiński G. (ed.) Integration as Solution for Advanced Smart Urban Transport Systems. Advances in Intelligent Systems and Computing, vol. 844, pp. 161–172 (2019)
11. Macioszek, E., Lach, D.: Analysis of the results of general traffic measurements in the west Pomeranian Voivodeship from 2005 to 2015. Sci. J. Silesian Univ. Technol. Ser. Transp. **97**, 93–104 (2017)
12. Pijoan, A., Kamara-Esteban, O., Alonso-Vicario, A., Borges, C.E.: Transport choice modeling for the evaluation of new transport policies. Sustainability **10**, 1230 (2018)
13. Staniek, M.: Stereo vision method application to road inspection. Baltic J. Road and Bridge Eng. **12**, 37–47 (2017)
14. Staniek, M.: Detection of cracks in asphalt pavement during road inspection processes. Sci. J. Silesian Univ. Technol. Ser. Transp. **96**, 175–184 (2017)

15. Małecki, K.: The importance of automatic traffic lights time algorithms to reduce the negative impact of transport on the urban environment. Transp. Res. Procedia **16**, 329–342 (2016)
16. Lejda, K., Mądziel, M., Siedlecka, S., Zielińska, E.: The future of public transport in light of solutions for sustainable transport development. Sci. J. Silesian Univ. Technol. Ser. Transp. **95**, 97–108 (2017)
17. Ocicka, B., Wieteska, G.: Sharing economy in logistics and supply chain management. LogForum **13**(2), 183–193 (2017)
18. Watróbski, J., Małecki, K., Kijewska, K., Iwan, S., Karczmarczyk, A., Thompson, R.G.: Multi-criteria analysis of electric vans for city logistics. Sustainability **9**(8), 1453 (2017)
19. Celiński, I.: Using GT Planner to improve the functioning of public transport. Recent advances in traffic engineering for transport networks and systems. In: Macioszek, E., Sierpiński, G. (eds.) Recent Advances in Traffic Engineering for Transport Networks and Systems. Lecture Notes in Networks and Systems, vol. 21, pp. 151–160 (2018)
20. Cárdenas, O., Valencia, A., Montt, C.: Congestion minimization through sustainable traffic management, A micro-simulation Approach. LogForum **14**(1), 21–31 (2018)
21. Zak, J., Galinska, B.: Design and evaluation of global freight transportation solutions (Corridors), analysis of a real world case study. Transp. Res. Procedia **30**, 350–362 (2018)
22. Turoń, K., Czech, P., Juzek, M.: The concept of a walkable city as an alternative form of urban mobility. Sci. J. Silesian Univ. Technol. Ser. Transp. **95**, 223–230 (2017)
23. MetroBike, LLC, Number of bike-sharing programs worldwide. http://bike-sharing.blogspot.com/2018/
24. BikeSharing Map, Numer of bikes available in bike-sharing programs worldwide. www.bikesharingmap.com
25. Hamari, J., Sjöklint, M., Ukkonen, A.: The sharing economy: why people participate in collaborative consumption. J. Assoc. Inf. Sci. Technol. **67**(9), 2047–2059 (2016)
26. Bieliński, T., Ważna, A.: Hybridizing bike-sharing systems: the way to improve mobility in smart cities. Res. J. Univ. Gdańsk Transp. Econ. Logistics Transp. Smart Cities **79**, 53–63 (2018)
27. Bielinski, T., Wazna, A.: New generation of bike-sharing systems in china: lessons for european cities. J. Manage. Fin. Sci. **33**, 25–42 (2018)
28. Chen, F., Turon, K., Kłos, M., Czech, P., Pamuła, W., Sierpinski, G.: Fifth generation bikesharing systems: examples from Poland and China. Sci. J. Silesian Univ. Technol. Ser. Transp. **99**, 5–13 (2018)
29. Midgley, P.: Bicycle-sharing schemes: enhancing sustainable mobility in urban areas, Background Paper No. 8 CSD19/2011/BP8: 1–26. New York: United Nations Department of Economic and Social Affairs. https://sustainabledevelopment.un.org/content/dsd/resources/res_pdfs/csd-19/Background-Paper8-P.Midgley-Bicycle.pdf
30. Shaheen, S., Stacey, G., Hua, Z.: Bikesharing in Europe, the Americas, and Asia: past, present, and future. Transp. Res. Rec.: J. Transp. Res. Board **2143**, 1–20 (2010)
31. Tóth, J., Mátrai, T.: Comparative assessment of public bike sharing systems. Transp. Res. Procedia **14**, 2344–2351 (2016)
32. Polish Network Portal, Bike Sharing, Five Generations Later: What's Next?. https://www.polisnetwork.eu/publicdocuments/download/2224/document/3c_brink.pdf
33. Macioszek, E., Lach, D.: Analysis of traffic conditions at the brzezinska and nowochrzanowska intersection in Myslowice (Silesian Province, Poland). Sci. J. Silesian Univ. Technol. Ser. Transp. **98**, 81–88 (2018)
34. Macioszek, E., Lach, D.: Comparative analysis of the results of general traffic measurements for the Silesian Voivodeship and Poland. Sci. J. Silesian Univ. Technol. Ser. Transp. **100**, 105–113 (2018)

35. Staniek, M.: Road pavement condition as a determinant of travelling comfort. In: Sierpiński, G. (ed.) Intelligent Transport Systems and Travel Behaviour. Advances in Intelligent Systems and Computing, vol. 505, pp. 99–107. Springer (2017)
36. Pucher, J., Buehler, R.: At the frontiers of cycling: policy innovations in the Netherlands, Denmark, and Germany. World Transp. Pol. Pract. **13**, 9–56 (2007)
37. Pucher, J., Buehler, R.: Cycling towards a more sustainable transport future. Transp. Rev. **37**(6), 689–694 (2017)
38. Pucher, J., Buehler, R.: Making cycling irresistible: Lessons from the Netherlands, Denmark and Germany. Transp. Rev. **28**(4), 495–528 (2008)

Finding the Way at Kraków Główny Railway Station: Preliminary Eye Tracker Experiment

Anton Pashkevich[1(✉)], Eduard Bairamov[2], Tomasz E. Burghardt[3], and Matus Sucha[1]

[1] Palacký University Olomouc, Olomouc, Czech Republic
{anton.pashkevich,matus.sucha}@upol.cz
[2] Cracow University of Technology, Kraków, Poland
edouard.bairamov@gmail.com
[3] M. Swarovski GmbH, Amstetten, Austria
tomasz.burghardt@swarco.com

Abstract. Eye tracker was used to study 25 young people from three nations who were given the task of finding their way to the main railway station in Kraków, Poland (station located underground, with access through a shopping mall), locating a ticket counter there, and going to a platform. Analysed were the paths selected by the test participants to reach these check points and their gazes at various directional signs. While the shortest distance for the task was 339 m, average participant travelled 503 m and the longest route was 1026 m. Observed were five confusion points, when most of the participants had to stop or turn around to find a sign pointing in the desired direction. There was a substantial confusion during the search for ticket counter in the station hall. The results demonstrate the weakness of the signage location and clarity. The outcome can be used by engineers conceiving such transportation hubs as a tool to optimize signage design.

Keywords: Eye tracker · Pedestrian · Transportation hub · Kraków · Wayfinding

1 Background

1.1 Introduction

Whereas in the past railway stations were prominent free-standing buildings, now they are quite frequently located as separate sections of shopping areas. In Kraków, Poland the main railway station building, which opened in 1847, was vacated from its role in 2014 while the station function was moved underground. The main pedestrian access to the new station from the side of the Old Town is through a shopping mall, which might cause misperception about the route amongst passengers not familiar with the design. The station hall is located directly under the platforms (10 tracks on 5 platforms and 2 bypass tracks) and above a tram stop in a tunnel and has a total of 10 levels. In 2017, approximately 16 million train passengers used the station, which served about 285 train departures daily [1]. Since Kraków is visited annually by 3.1 million of foreign tourists [2], some of whom can travel by train to both local tourist-frequented destinations (like Wieliczka salt mine, former German concentration camp Auschwitz, or Kraków airport) or to long-distance destinations in Poland and abroad, clear passenger information that can be comprehended by people from various cultures is demanded. Majority of the directional signs at the station are pictograms; they are in most places supplemented with descriptions in Polish (larger font) and English (smaller font).

Wayfinding is a commonly studied topic, with various aims and outcomes [3]. However, the use of mobile eye tracker in a non-simulated environment is a novelty amongst wayfinding research. To the best of our knowledge this is one of the first reports of such assessments and the pioneering study of the new Kraków Główny railway station. The goal of the presented work was evaluation of the route selection and signage permitting for orienteering at the railway station by a casual young person. Path selection and glances at various signs were compared to assess their usefulness. Measured were the number of confusions, distance travelled, and time to find the appropriate check points. In this article are presented the preliminary results, comprising the averages of distance walked, time to look at various signs, and path selection.

1.2 Eye Tracking

Eye tracking is a well-known research technique dating back to the beginning of 20th century. It is a process to measure the eye movement that permits to identify the points of gazes. Eye tracking measurement device allows to record these types of information. Currently, there are four large groups of eye trackers: scleral contact lens/search coil, electro-oculography (EOG), photo- or video-oculography (POG or VOG), and video-based combined pupil and corneal reflection. Each one has its own limitation as well as advantages and disadvantages [4, 5].

In the past two decades, there is an interest in using eye tracking tools to investigate navigation tasks, but the published literature still remains limited. The differences between males and females in reading of maps revealed no differences between sexes, but dissimilarities in giving directions based on the viewed maps indicated different cognitive processes [6]. Somewhat related work of directing and navigating in a complex indoor environment combined with assessment of the difference between attitudes and behaviour by realising a wayfinding tasks was reported several years later [7, 8]. These and similar studies were done mostly under laboratory conditions. More recently, with development of portable eye tracking devices, analyses in the field commenced. The difference between the objects that people saw during experiments in a busy urban environment and which they remember afterwards was reported [9]. More complicated field navigation experiment, including self-orientation, target search in the environment and on the map, route memorization and destination achievement, was recently reported [10]. Interesting research work in context of this article was a study of Vienna's main railway station done in 2017, where the usage of virtual environment is presented as a new approach to analyse a visual attention inside infrastructural objects [11].

2 Experiment Design

2.1 Participants

Test participants were three groups of students: from Russia, from Croatia, and from Poland (including a few from Kraków, who could be considered as being a control group, familiar with the railway station). All of the participants were volunteers, not compensated for their effort. Prior to starting the experiment, each of the participants filled a questionnaire with basic demographic and other questions (provided in Table 1). They all claimed to have uncorrected or corrected 6/6 vision. Self-assessment of English language knowledge was an important consideration because of some signs using this language in addition to pictograms. Ethical guidelines set by Politechnika Krakowska and by Palacký University Olomouc were followed at all times.

Table 1. Test participants.

Country	Total		Russia		Croatia		Poland	
Sex	Male	Female	Male	Female	Male	Female	Male	Female
Number	11	14	1	4	3	5	7	5
Age [years]	22.5 (5.2)	23.8 (7.0)	21	24.8 (9.5)	27.3 (9.3)	25.6 (8.1)	20.7 (0.8)	21.0 (0.7)
English[a]	–	–	2	3.3 (0.5)	3.0 (0.0)	3.2 (0.4)	3.6 (0.5)	3.6 (0.5)

Standard deviations are provided in parentheses. [a]Self-assessed knowledge of English language, on a 1–5 scale, with 1 being poor, 3 – average and sufficient to comprehend all written signs, and 5 being perfect knowledge.

2.2 Equipment and Data Processing

Data collection was accomplished with eye tracking spectacles Tobii Pro Glasses 2 (Tobii AB, Danderyd, Sweden) and processed with Tobii Pro Lab. The selected eye-tracker, belonging to the VOG group, allows for recording of typical eye-tracking parameters: gaze, fixation, fixation duration, and saccade. Equipment was calibrated for each of the test participant with Tobii Pro Glasses Controller software.

2.3 Test

For each participant, the starting point was at a bus stop located across the street from the shopping mall building, through which one enters the train station. The participants were instructed in the following manner (the instructions were in English, except participants from Poland, for whom Polish language was used): "Go from here to the railway station, find the departures schedule, then walk to the ticket counter, and at the end enter platform 4." The eye tracker performance was verified and the participants were allowed to find their way. Test assistants were constantly monitoring the selected route, noting confusion points, etc. In many times, the test participants requested help or confirmation of reaching the checkpoint from the assistants, to which a minimal response was given. Upon completion of the task, the participants were given another brief questionnaire, asking for their assessment of the difficulties to locate the check points. The studies were done during daytime on work days, with average passenger and shopping mall visitors' traffic.

3 Results

3.1 Route Selection

The shortest possible route to complete the task, based on the analysis done by the authors, was 339 m, but it should be noted that during the day it is not necessarily the fastest route. The route length selected by the participants varied from 396 to 1026 m, with average 503 m (for foreigners, average 557 m). The results, broken into the countries of origin are given in Table 2. Definitely, the longest route was taken by participants from Croatia, on average 602 m as compared to 485 m for participants from Russia and 445 m for those from Poland. Similarly, participants from Croatia were the slowest and were the most confused and even lost. Due to different standard paces of different people [12], it has to be noted that the time and speed are not necessarily fully representative; nonetheless, combined with the number of confusions and the distance travelled, they give an indication of the difficulties in locating the target points. The test participants from Croatia and Russia in majority of cases were requesting a confirmation from the assistants that a checkpoint was reached and occasionally requesting help in proceeding. Response from the test assistants was given only on some occasions, as deemed necessary. Half of the test participants from Croatia became lost (went to another platform or to the tram stop and were waiting there, without realisation of not completing the tasks). A few of the test participants incorrectly considered ticket vending machines as ticket counters. List of major inadequacies is shown in Table 2.

Table 2. Routes taken by test participants and their confusion.

Task	Average	Minimum	Maximum	Russia	Croatia	Poland
Route length [m]	503 (134)	396	1026	485 (46)	602 (203)	445 (38)
Time to complete [s]	484 (146)	267	754	484 (74)	647 (94)	375 (82)
Speed [m/s]	1.1 (0.23)	0.6	1.5	1.0 (0.15)	0.9 (0.24)	1.2 (0.18)
Confusions	2.0 (1.94)	0	6	1.6 (0.89)	4.1 (1.46)	0.8 (1.22)
Request confirmation from assistants				3 (60%)	5 (63%)	0
Received confirmation from assistants				1 (20%)	4 (50%)	0
Considered ticket vending machine as ticket counters				2 (40%)	5 (63%)	5 (42%)
Lost (had to be directed)				0	4 (50%)	0

Standard deviations given in parentheses. Percentages pertain to the test participants from a particular country.

3.2 Route Visualisation and Sign Identification

For an easy data presentation, a visualisation of the paths taken by the test participants, containing visual information signs, was prepared. A systematic method for identification of various signs was developed for the use in this and subsequent analyses in such transport hub environments. The system comprises location of the sign, sign category, and sign number, according to the coding given below, which was arbitrarily allocated. In case of signs combining different categories, all of them are shown (because sign location within a level is only one letter, subsequent letters signify various sign categories and do not need to be separated); signs visible from different levels are marked as belonging to the level at which they are located. Double-sided signs are considered as two separate signs.

- First digit – level number (starting from level 0 as lowest in the tested environment)
  - In our case, level 0 would be the underground tram stop, ticket counters are located at level 2, station hall and lower floor of the shopping mall are level 3, additional pedestrian tunnel is at level 4, station mezzanine at level 5, ground level at the side of Old Town and the ground level of the shopping mall are at level 6, ground level at the other side of the station is at level 7, platforms are located at level 8, and the overhead car park occupies level 10
- First letter – sign location
  - A – before entrance to the transportation hub region (outside of the building)
  - B – inside a building, but not inside an area being the transport hub itself
  - C – within the transportation hub, but not at a hall or platforms
  - D – within the transportation hub hall
    - In the case of this article and station set-up, ticket counters are assumed as belonging to the station hall
  - E – at platforms
- Second (and subsequent) letter – sign category
  - Z – directing to transportation hub
  - Y – static schedule
  - X – dynamic schedule
  - W – directing to an exit from transportation hub (to city, to shopping mall, and other 'way out' signs)
  - V – directing to transportation hub elements (to ticket counters, to waiting hall, to information, etc.)
  - U – directing to departure platforms (including train, tram, bus, etc.)
  - T – informing of location within the transportation hub (station hall, waiting room, etc.)
  - S – arrival and departure signs from a particular platform
  - R – platform sign (platform number, platform sector, track number, station name, etc.)

- P – various plaques (no smoking, lift, warning: escalator, no entry, etc.)
- O – emergency exit signs
- N – maps and schemes of the transportation hub
- M – maps of the town, station area, and regional maps
- L – other signs
- K – signs for handicapped people (3D guides, Braille signs, etc.)
* Digits – number of a particular sign
* * - symbolises a sign that was observed

The overlay of selected routes is provided on schematic drawings, not to scale and depicting only the features noted in this article. The legend shown on Fig. 1 applies to other schemes, too. The paths that the participants took at level 6 (street and upper level of the shopping mall) is presented in Fig. 1, their paths at the pedestrian tunnel (level 4) and lower floor of the shopping mall (level 3) are given in Fig. 2 and in the station hall area (station hall at level 3, ticket counters at level 2, and mezzanine at level 5) in Fig. 3. Each of the shown paths was coded by colour based on the participant's nationality (blue dotted lines – Russia, green dashed lines – Croatia, and red solid lines – Poland). Location of signs mentioned in the text is shown on the schemes. Because some of the test participants who confused ticket counters with ticket vending machines, these were also marked.

It can be clearly seen in Figs. 1 and 2 that about half of the participants selected the way through the shopping mall, either through its upper or lower level, whereas the other half preferred walking outside until the main entrance. Two participants from Croatia apparently already became confused and selected a circuitous, but still a reasonable route; however, further, they seemed to have been lost and entered the station hall through a platform. For one polish participant, who was familiar enough with the station layout, selection of such a route may be considered as based on purely personal choice. The major conundrum location for almost all of the test participants was at the entrance from the shopping mall to the station hall (at level 3), as evidenced by search for visual information and walking in circles, shown in Fig. 3. Subsequently, several routes were chosen to the ticket counters. The test participants also selected different entrances to platform 4 (one of the two stairways or one of the two lifts), but without major recorded confusions.

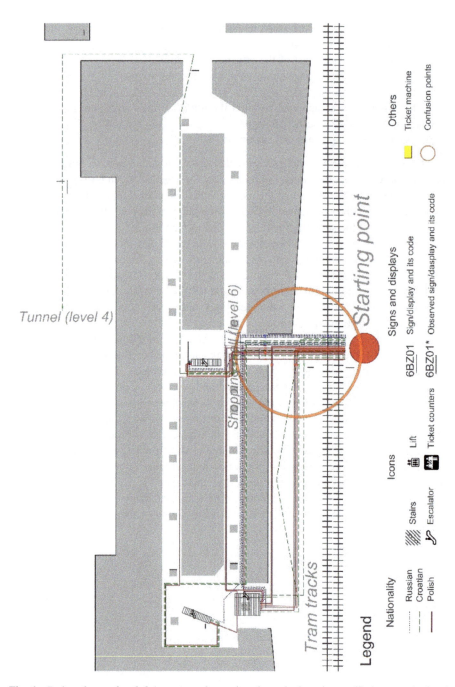

Fig. 1. Paths taken at level 6 (access to the station through shopping mall) (source: Authors)

Finding the Way at Kraków Główny Railway Station 245

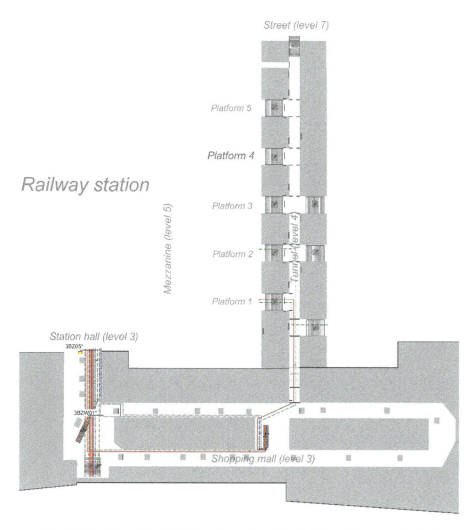

Fig. 2. Paths taken at level 3 (shopping mall) and level 4 (tunnel) (source: Authors)

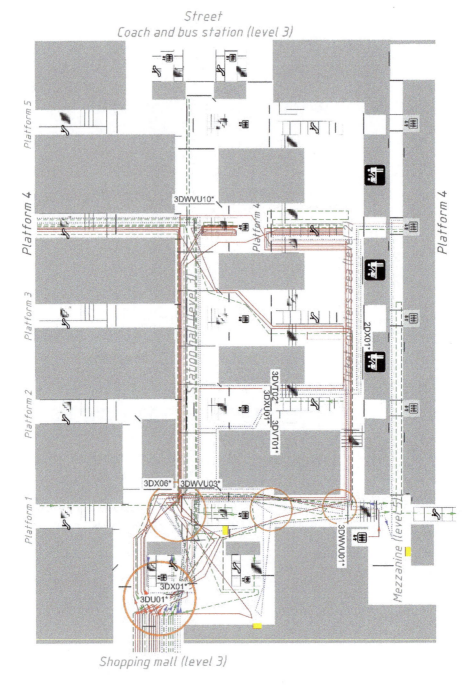

Fig. 3. Paths taken at the station hall (level 3), including the ticket counters area (level 2), and at the mezzanine (level 5) (source: Authors)

3.3 Signs Observation

At the area where the test participants travelled, the authors counted 193 signs that could be have been seen by them. Out of these signs, 96 (50%) were observed by at least one test participant. For analysis, the signs are divided herein into those directing to station entrance (sign category Z), schedule signs (signs category X and Y), and navigation signs to find the ticket counters and platforms (sign category V, U, and T, and S). Without discussion are left 3 maps and schemes (sign category N), which were observed by 4 (16%) of test participants; quite surprisingly, in no case they helped in wayfinding.

All of the pictures shown below were taken on a separate occasion by the authors. Observation distance is the average distance, from which a sign was observed (note that a sign could be observed more than once by the same test participant), and observation time was the average length of fixation per observation. For the 'other signs', average number of observers is given. Standard deviations are provided in parentheses.

Z – Signs Directing to Station Entrance

Amongst 21 signs directing to the train station, located before and inside the shopping mall, at levels 3 and 6, observed were 17 (81%). Two of them were observed by 15–17 (60–68%) participants, from an average distance 16.3–17.6 m, and the observation time was 2.1–2.5 s. The remaining fifteen signs were noticed by 1–11 (4–44%) participants from an average distance 13.0 m and mean observation time 1.9 s. At present it is not possible to determine whether the most frequent observations were caused by the signs' dimensions or location. Selected signs are shown in Table 3; amongst the frequently observed, lacking is any sign outside of the station building.

Table 3. Most frequently observed signs directing to station entrance (category Z).

Sign picture[a]	Sign number and dimensions [m]	Observers	Observation Distance [m]	Time [s]
	3BZW01* 1.0 × 5.5	15	16.3 (7.9)	2.51 (2.36)
	3BZ05* 0.4 × 11.0	17	17.6 (9.7)	2.14 (1.39)
All other 15 signs category Z that were observed	0.3–1.0 × 1.0–4.0	3 (3)	13.0 (9.8)	1.90 (2.5)

X and Y – Schedule Signs

Amongst 56 train departure schedule signs that were visible to the test participants, observed were only 10 (18%): 8 dynamic (sign category X) and 2 static (sign category Y). From those, two monitor displays were read by 14 (56%) test participants. One of these displays is quite large (2.0 × 5.5 m), whereas another one is small (0.4 × 0.6 m). The large display was observed for an average of 3.3 s (range 0.1–11.2 s). This schedule display can be quite confusing because both arrivals and departures are given on the same background and using the same layout. There are also another two similar schedule displays near ticket counters. Surprisingly, the rest of train schedule signs, which include 15 small departure monitors (0.4 × 0.6 m) and 23 separate train departure signs for different platforms (0.2 × 1.0 m), did not attract any attention of test participants. The departures schedule as a checkpoint shall be discussed separately. The observed schedule signs, along with their dimensions, average observation distances and times are shown in Table 4.

Table 4. Observed train schedule signs (categories X and Y).

Sign picture[a]	Sign number and dimensions [m]	Observers	Observation	
			Distance [m]	Time [s]
	3DX06* 2.0 × 5.5 (display and overhead title)	14	15.9 (12.3)	3.27 (3.07)
	3DX01* 0.4 × 0.6 (display only)	14	10.1 (5.2)	2.20 (2.12)
	2DX01* 2.0 × 5.5 (display and overhead title)	6	15.6 (9.6)	3.46 (3.11)
All other 7 train schedules (sign categories X and Y)	0.2–2.0 × 0.6–5.5	1 (0)	5.6 (3.4)	5.26 (4.93)

S, T, U, and V – Directional Signs

There were counted 107 signs categories S, T, U, and V that were visible for participants in the station area (levels 0, 1, 4, 5 and 6, location C; levels 2 and 3, location D; level 8, location E). Amongst them, 66 (62%) were observed by at least one person. None of signs category S was observed, which might have been caused by the lack of specification of a particular train during the instructions. Most frequently observed were 4 signs, which gained attention of 12–16 (48–64%) test participants; their average observation was from the distance of 12.2–19.2 m and the total fixation time per participant varied from 1.09 to 2.98 s. Sign 3DWVU01*, located at level 3, next to an entrance to a mezzanine and a passage to level 2, was an outlier in terms of fixation time, which varied from 0.42 to 7.4 s. It is most likely caused by its complication. This sign is also appearing incorrect and unclear because it points straight to stairs whereas the sign shows both stairs and an escalator. Arrows pointing up and left/right are not straightforward, either. Incorrect is also sign 3DU01*, at the entrance to the tram tunnel, displaying both stairway and lift whereas it points directly to an escalator. Information regarding the tram routes and directions is missing from this area, which is an inadequacy because there are two tram platforms with separate entrances.

Sign 3DU01* is the most prominent sign visible for anyone entering the station hall from the side of the shopping mall, at level 3, but it does not direct to the station hall. Selected most frequently observed directional signs are shown in Table 5.

Table 5. Most frequently observed directional signs (categories S, T, U, and V).

Sign picture	Sign number and dimensions [m]	Observers	Observation	
			Distance [m]	Time [s]
	3DWVU01* 1.5 × 5.5	12	17.3 (9.3)	2.98 (1.84)
	3DWVU03*[a] 0.4 × 7.5	16	19.2 (8.8)	2.44 (1.56)
	3DWVU10* 0.4 × 7.5	13	17.5 (5.8)	1.80 (2.32)
	3DU01* 0.5 × 2.5	16	12.2 (3.2)	1.09 (0.80)
All other 62 signs category T, U, and V that were observed within levels 0, 1, 4, 5 and 6, location C; levels 2 and 3, location D; and level 8, location E	0.2–1.5 × 0.5–7.5	3 (2)	13.6 (7.6)	1.59 (1.44)

[a] Subsequent analysis revealed issues with processing software and positioning of this sign, which was resolved into signs 3DWVU03* and 3DWVU05* [13].

3.4 Confusion Points and Comments from the Participants

Confusion points are marked with large circles on Figs. 1 and 3. They were mostly at the entrance to the station hall and can be associated with the search for ticket counters, which are accessible via various routes, but themselves are not visible to anyone entering the station. This can be noted as a major inadequacy of this railway station design.

All of the test participants left brief comments; generally positive, but there were repeated two major complaints amongst all participants that can be summarised:

- Where is the station? There are no signs anywhere at the bus stop where the test starts.
- The way to ticket counters resembles a maze.

4 Discussion and Conclusion

The results consistently show test participants from Croatia as being the most confused, which translated into walking the longest routes, getting lost, and needing most time. Whereas this could be a result of them being foreigners, it must be noted that test participants from Russia selected routes only marginally longer than local inhabitants. Amongst possible reasons for better wayfinding skills that was measured amongst the participants from Russia as compared to those from Croatia one could list for example differences and habits, such as the use of intercity public transportation. In the questionnaires, which were made as simple as possible, there was no space for such enquiries. The extent, to which the test participants required confirmations and directions from the research assistants seems to indicate not only inadequate information found at the railway station, but also the lack of confidence in the visual input. None of the test participants, even at the confusion point, used station layout schemes, which may indicate either their lack of map reading skills or the trust in signage hub.

Amongst the deficiencies of signage and station layout one must point out the design of the area where the ticket counters are located. They are hidden from the view of travellers entering the station (Cf. station hall layout shown in Fig. 3) and there are numerous possible ways to access them. Hence, one could simultaneously see several directing signs, which can create confusion, indeed. In addition, some signs appear to be incorrectly (or unfortunately) located, as one can see on an exemplary signage shown in Fig. 4 (signs from left to right 3DVT02*, 3DXU01* and 3DVT01*). The stairs symbol at the right-side sign 3DVT01* is above a ramp whereas a wheelchair symbol on the left-side sign 3DVT02* is directing to stairs; the information and shower symbols might indicate that the ramp and stairs are not going to the same lower area (level 2); the wheelchair symbol at the middle sign, next to the exit symbol, is not indicating that it actually points to a lift and the arrow might suggest that one needs not to turn around, but make two left turns. Proposing a way to improve the signage is beyond the scope of this preliminary assessment.

Additionally, very inconvenient sets of stairways at the approaches to the ticket counters (located at level 2, whereas the station hall is 1.5 m higher, at level 3) must be noted. These stairways do not appear to be making any sense to a casual observer and user of the facilities. The concept of making the main underground station hall at two levels appears inefficient and is a barrier for people unable to use stairs and a difficulty for travellers carrying heavy luggage. There must be noted that a ramp like shown in Fig. 4 is only next to platforms number 2 and 5; near the other three platforms there are only stairs.

Fig. 4. Exemplary signage that might be confusing

A very significant limitation of this evaluation was the exclusion of older population. Elderly may view the directional signs differently, because of previous experiences and/or habituation [14, 15]. It is likely that an experienced older traveller who had seen many station designs would avoid any confusions whereas elderly occasional traveller could become completely lost. Another deficiency of the presented work was the selection of the starting point at a bus stop, which, it was realised too late, was quite unlikely to be used by many tourists.

Future work that is planned on the same area includes finding a path that would exclude using any stairways, which are highly uncomfortable for people with suitcases, those using a wheelchair, or travellers with a pram and evaluation of other access routes to the station hall. It is also planned to evaluate incoming passengers and their routes to various public transportation facilities (there are five groups of tram and bus stops that serve the railway station). Assessment of signage solutions used at other stations and in other countries is also proposed. Interesting side work would be checking the passengers' behaviour in response to an unexpected platform or track change or the need to use an ersatz transportation mode. Additional studies that were beyond the scope of this preliminary evaluation, such as including or excluding the 'noise' caused by the congestion, can also be performed.

In conclusion, the results from this field experiment demonstrated that the design of the new Kraków Główny railway station and directional signs that should guide passengers are at most satisfactory, very far from optimum. Despite a plethora of directional signs, they are not channelling travellers using the simplest way. Some signs are confusing and erroneous. Particular difficulty was noted in finding the ticket counters, which are hidden from the view of people entering the underground station.

References

1. Największe stacje kolejowe w Polsce. Urząd Transportu Kolejowego. Warszawa (2018). [in Polish]. https://www.utk.gov.pl/pl/raporty-i-analizy/analizy-i-monitoring/analizy-i-opracowania/14508,Najwieksze-stacje-kolejowe-w-Polsce.html
2. Ruch turystyczny w Krakowie w roku 2018. Urząd Miasta Krakowa, prezentacja. Kraków (2018). [in Polish] https://www.bip.krakow.pl/plik.php?zid=226115
3. Farr, A.C., Kleinschmidt, T., Yarlagadda, P., Mengersen, K.: Wayfinding: a simple concept, a complex process. Transp. Rev. **32**(6), 715–743 (2012). https://doi.org/10.1080/01441647.2012.712555
4. Duchowski, A.T.: Eye Tracking Methodology: Theory and Practice. Springer, Cham (2017). https://doi.org/10.1007/978-3-319-57883-5_5
5. Navalpakkam, V., Churchill, E.F.: Eye tracking: a brief introduction. In: Olson, J.S., Kellogg, W.A. (eds.) Ways of Knowing in HCI, pp. 323–348. Springer, New York (2014). https://doi.org/10.1007/978-1-4939-0378-8_13
6. MacFadden, A., Elias, L., Saucier, D.: Males and females scan maps similarly, but give directions differently. Brain Cogn. **53**(2), 297–300 (2003). https://doi.org/10.1016/S0278-2626(03)00130-1
7. Hund, A.M., Padgitt, A.J.: Direction giving and following in the service of wayfinding in a complex indoor environment. J. Environ. Psychol. **30**(4), 553–564 (2010). https://doi.org/10.1016/j.jenvp.2010.01.002
8. Hund, A.M., Schmettow, M., Noordzij, M.L.: The impact of culture and recipient perspective on direction giving in the service of wayfinding. J. Environ. Psychol. **32**(4), 327–336 (2012). https://doi.org/10.1016/j.jenvp.2012.05.007
9. Sayegh, A., Andreani, S., Li, L., Rudin, J., Yan, X.: A new method for urban spatial analysis: measuring gaze, attention, and memory in the built environment. In: Proceedings of the 1st International ACM SIGSPATIAL Workshop on Smart Cities and Urban Analytics; Seattle, WA, 3–6 November 2015, pp. 42–46 (2015)
10. Liao, H., Dong, W., Huang, H., Gartner, G., Liu, H.: Inferring user tasks in pedestrian navigation from eye movement data in real-world environments. Int. J. Geograph. Inform. Sci. **33**(4), 739–763 (2019). https://doi.org/10.1080/13658816.2018.1482554
11. Schrom-Feiertag, H., Settgast, V., Seer, S.: Evaluation of indoor guidance systems using eye tracking in an immersive virtual environment. Spat. Cogn. Comput. **17**(1–2), 163–183 (2017). https://doi.org/10.1080/13875868.2016.1228654
12. Levine, R.V., Norenzayan, A.: The pace of life in 31 countries. J. Cross Cult. Psychol. **30**(2), 178–205 (1999). https://doi.org/10.1177/0022022199030002003

13. Pashkevich, A., Bairamov, E., Burghardt, T. E., Lenik, P., Sucha, M.: Finding the way at Kraków Główny railway station: directional signs solving confusion in eye tracker study. At 23rd International Scientific Conference Transport Means 2019; Palanga, Lithuania, 2–4 October 2019 (manuscript submitted)
14. Underwood, G., Phelps, N., Wright, C., Van Loon, E., Galpin, A.: Eye fixation scanpaths of younger and older drivers in a hazard perception task. Ophthal. Physiol. Opt. **25**(4), 346–356 (2005). https://doi.org/10.1111/j.1475-1313.2005.00290.x
15. Fancello, G., Pinna, C., Fadda, P.: Visual perception of the roundabout in old age. WIT Trans. Built Environ. **130**, 721–732 (2013). https://doi.org/10.2495/UT130581

Using the Kalman Filter for Purposes of Road Condition Assessment

Marcin Staniek[✉]

Faculty of Transport, Silesian University of Technology, Katowice, Poland
marcin.staniek@polsl.pl

Abstract. This article provides a discussion on the use of the Kalman filter at the stage of analysis of vehicle motion dynamics to assess the condition of the road transport infrastructure. The tool used to record the vehicle motion dynamics was the Road Condition Tool application designed and implemented as a part of the international S-mile project under the programme entitled ERA-NET: Transport – Sustainable logistics and supply chains. An outcome of the data recording and analysis is a description of the operating condition of road pavements. The concept addressed in the study, namely the implementation of the Kalman filter at the data analysis stage, makes it possible to reduce the input data set, which affects the measuring tool's response time. On the other hand, the adopted signal filtration procedure allows for the infrastructure condition to be described in a more favourable manner from the perspective of the conclusions to be formulated.

Keywords: Road pavement condition · Assessment of road infrastructure · Vehicle motion dynamics · Smart city · ICT application

1 Introduction

Assessment of the condition of roads in the network stems from the process of road pavement diagnostics covering the stages of acquisition, accumulation and processing of information concerning the service properties of roads. Up-to-date, objective and accurate data are crucial for maintaining the road network operational. This is why access to information which proves reliable in both quantitative and qualitative terms is prerequisite to ensure reasonable maintenance of the road network.

Pavement Management Systems (PMS) [1] feature two independent processes: one related to identification of parameters of the road infrastructure condition description, and the other related to the process of analysis and assessment of the data acquired. The identification process must provide reliable information about the given structure's geometry and physical properties, i.e. the structure's load capacity characteristics, longitudinal and transverse profile, as well as detection and parameterisation of defects. The process of analysis and assessment of the pavement condition relies on the data previously acquired and then confronted with historical data, and it uses predictive models to anticipate degrading of the given structure. The scope and type of the road condition analysis as well as the methods used to assess it are conditional on specific needs and goals of the road infrastructure management policy [2]. For transport infrastructure, in addition to the condition assessment, it is necessary to assess the traffic conditions as well as road occupancy [3–6].

2 Literature Review

A study of the available literature pertaining to the assessment of the condition of roads highlights some specialised tools used by infrastructure administrators to monitor operating parameters. They include commonly available technologies of image processing and recognition [7], but also neural networks are in use [8]. What is becoming increasingly typical is the use of solutions based on assessment of vehicle motion dynamics with reference to linear accelerations resulting from the movement of vehicles in a road network of specific technical condition. One of popular approaches often addressed in the literature is based on a multi-channel data logger unit recording linear accelerations induced by vibrations of the driven wheel moving on the road section subject to examination, while at the same time it is loaded by the mass applied. The fact that recorded linear accelerations (taking average speed into account) depend on the road pavement unevenness has been confirmed by studies whose results are provided in paper [9]. It has also been proved that data provided by the MEMS sensors are most accurate in describing the road pavement condition in the measuring frequency range from 40 to 50 Hz.

Publication [10] describes an architecture of a system which comprises a data logger, a communication module installed on-board the test vehicle and an external data processing system. Field tests of the prototype system were conducted in the Chinese province of Zhejiang, while the results thus obtained imply that, compared to the method of unevenness assessment by means of laser tools, the relative error of the system in question is below 10%, which confirms the high accuracy, efficiency and reliability of the solution proposed. What proves to be crucial in the analysis of road description signals is filtering out the noise emerging as an outcome of the vehicle mass system damping. In the process of road unevenness identification, one typically uses half- or quarter-vehicle models of the suspension system [11, 12].

Paper [13] proposes to implement a transfer function which provides a description of running over a speed bump of pre-defined shape and dimensions in order to determine the motion parameters of a vehicle recording linear accelerations. The proposed transfer function, established in the process of simulation of a half-vehicle suspension system model, reflects the vehicle tilt as well as the measuring device's installation point.

An assessment of the condition of pedestrian pathways and bicycle roads by application of a method which consists in measuring the International Roughness Index (IRI) has been discussed in publication [14]. What has also been proposed in that paper is some algorithms enabling recognition of road potholes and bumps. The research in question, including an assessment of the method proposed, confirms that the IRI values measured by application of the said method were highly and positively correlated with values delivered by measurements conducted by means of measuring tools typically used by road infrastructure administrators. Furthermore, the proposed algorithm was capable of identifying potholes and bumps directly run over by a bicycle.

Paper [15] provides an example of how neural networks can be used to assess the condition of road pavement based on linear accelerations recorded with an accelerometer. The research results imply that the implemented neural network is capable of reconstructing both the profile and the typical defects of roads, thus ensuring a high level of accuracy in mapping the description of a structure's condition. The assessment methodology is based on an assumption that real-life road profiles are unavailable while performing measurements, while data are acquired exactly by measurements conducted in the road network.

An analysis of vehicle motion dynamics determined by the road pavement condition in the domain of time and frequency as well as wavelet transformation has been provided in paper [16]. A method was proposed to reduce the impact of velocity, slope and drifts on the signals of description of motion dynamics. A wavelet decomposition analysis was performed for signal processing of inertial sensor signals and the Support Vector Machine (SVM). This approach enabled implementation of a road pavement defect classifier characterised by an accuracy of 90% for major road defects, regardless of the vehicle type and the defect location within a right-of-way. Furthermore, according to the solution in question, the condition of roads is assessed on a real-time basis. The acquired data using various measurement methods can be attributed to the Open Street Map [17].

In order to apply the Kalman filter [18] with reference to data of vehicle motion dynamics in a real-life road network, the author of this article used a method which consisted in recording and analysing linear acceleration signals, as well as the method's implementation under a solution known as Road Condition Tool (RCT). RCT is a component of the S-mileSys platform whose main goals include efficient transport route planning [19–31] over the first/last mile to support management of the supply chain's individual elements connected with the freight transport sector. The idea behind the S-mileSys platform is pursued under the international project entitled "Smart platform to integrate different freight transport means, manage and foster first and last mile in supply chains (S-mile)" as a part of the "Sustainable Logistics and Supply Chains" competition within the framework of the ERANET Transport III programme. The project was implemented by a consortium of six institutions, including businesses and higher technical schools from three countries: Spain, Turkey and Poland.

The implemented RCT tool is a mobile device, e.g. a tablet or a smartphone, running on the Android operating system and featuring the RCT Mobile application developed under the S-mile project as well as the RCT Server data processing module. The implementation of the RCT tool resulted from a review of the available solutions for identification of road traffic parameters and road pavement condition assessment as well as from an analysis of their comprehensiveness and simplicity of the measurement procedure involved [32]. This article describes how the RCT tool will be extended by the Kalman filter [33].

3 Filtering of Signals Describing Vehicle Motion Dynamics

The Kalman filter is a popular algorithm used to estimate the internal condition of an object based on measurements of the object's input and output. As per its basic rationale, the Kalman filter is applied to handle linear systems, while its modification makes it possible to work with systems that are non-stationary, variable in time and non-linear (Extended Kalman Filter) [33]. Its application makes the estimation of an object's condition statistically optimum. A discrete model of the process in the state space is given by the following relations (1) and (2):

$$x(t+1) = Ax(t) + Bu(t) + v(t) \qquad (1)$$

$$y(t) = Cx(t) + w(t) \qquad (2)$$

where:

x(t) – state of the object at instant t,
y(t) – object's output,
A, B, C – matrices of the object's state, input and output, respectively,
v(t), w(t) – process noise and measurement noise.

The Kalman filter's algorithm is divided into two stages. Stage one covers time updating, and it consists in calculating a one-step prediction of state, i.e. an a priori estimate (3), and its covariance matrix (4):

$$\hat{x}(t+1|t) = A\hat{x}(t|t) + Bu(t) \qquad (3)$$

$$P(t+1|t) = AP(t|t)A^T + V \qquad (4)$$

Another step is time incrementing, and the Kalman filter's algorithm proceeds to the updating of measurements which marks the beginning of the second stage characterised by the following set of equations:

$$\varepsilon(t) = y(t) - C\hat{x}(t|t-1) \qquad (5)$$

$$S(t) = CP(t|t-1)C^T + W \qquad (6)$$

$$K(t) = P(t|t-1)C^T S^{-1}(t) \qquad (7)$$

$$\hat{x}(t|t) = \hat{x}(t|t-1) + K(t)\varepsilon(t) \qquad (8)$$

$$P(t|t) = P(t|t-1) - K(t)S(t)K(t)^T \qquad (9)$$

where:

ε(t) – difference between last measurement y(t) and a value expected with reference to the current state estimation,
K(t) – Kalman gain matrix determining the measurement's effect on the state estimate from both a posteriori and a priori perspective.

According to the assumed road pavement condition identification model implemented in the RCT tool and described in several papers, including [33, 34], the Kalman algorithm was defined, intended to be used at the stage when differences in the vehicle's linear accelerations would be determined. The filtration algorithm's mathematical model in the prediction phase is described by relations (10) and (11).

$$x(t) = \hat{x}(t-1) \tag{10}$$

$$\bar{P}(t) = P(t-1) + V \tag{11}$$

The correlation phase, on the other hand, is given by relations (12), (13) and (14).

$$\mathbf{K}(t) = \bar{P}(t)(\bar{P}(t) + W)^{-1} \tag{12}$$

$$\hat{x}(t) = \hat{x}(t) + K(t)[z(t) - \hat{x}(t)] \tag{13}$$

$$P(t) = [1 - K(t)]\bar{P}(t) \tag{14}$$

Further figures provide a graphical interpretation of the signal filtration model proposed. Figure 1 shows a sample notation of the linear acceleration signal recorded using the RCT tool for a road of poor service condition, i.e. graded C [35], which requires planned repair works.

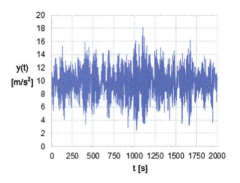

Fig. 1. Vertical linear accelerations recorded using the RCT tool

Another illustration, i.e. Fig. 2, shows standard deviations obtained from a set of differences between consecutive values of the acceleration signal in a pre-set calculation window for a chosen measurement section conforming with the notation provided in paper [32].

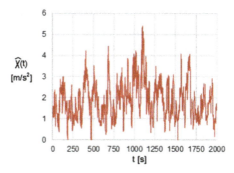

Fig. 2. Standard deviation of differences between consecutive measurements in a pre-set calculation window

The effect of Kalman filtering of a signal which describes changes in linear accelerations was illustrated on Fig. 3.

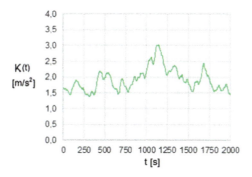

Fig. 3. Kalman filtering algorithm for a linear acceleration signal

4 Assessment of Efficiency of the Kalman Filtering Algorithm

What the author proposed as a means to assess the efficiency of the implemented filtering algorithm was to apply multi-class classifiers [36] making it possible to determine the affiliation of the analysed road section with pre-defined classes of the pavement condition description conforming with the PMS guidelines [35]. On such a basis, the errors made in classification to the pre-defined classes were analysed. As an effect thereof, a confusion matrix was developed, enabling assessment of the efficiency of the method proposed by determining values of the following indicators:

- number of correctly classified conditions of road pavement of the chosen class – TP (true positive),
- number of incorrectly classified conditions of road pavement of the chosen class – FN (false negative),

- number of classified conditions of road pavement not assigned to the chosen class (correctly rejected) – TN (true negative),
- number of classified conditions of road pavement incorrectly assigned to the chosen class – FP (false positive).

The measurement section for which the proposed method was assessed in terms of its functional efficiency comprised 215 actual cross-sections. The analysis of the errors made in the classification to the pre-defined classes of road condition description has been presented in the confusion matrix (Table 1).

Table 1. Confusion matrix for the Kalman filtering algorithm.

Operating classes of road pavement		Measurements with using Kalman filter			
		Class A	Class B	Class C	Class D
Reference measurements	Class A	77	6	1	0
	Class B	5	48	3	1
	Class C	0	4	41	2
	Class D	0	0	3	24

With reference to the confusion matrix, values of indicators TP, FN, FP and TN were determined. They have been collated in Table 2 for the assumed classes describing the condition of roads.

Table 2. Components of the measure of assessment of the Kalman filtering algorithm's efficiency.

Operating classes of road pavement	Measurements with using Kalman filter			
	TP	FN	FP	TN
Class A	77	7	5	113
Class B	48	9	10	142
Class C	41	6	7	149
Class D	24	3	3	166

Assuming the data provided in Table 2 and with reference to the relevant relationship as per [35], the following results have been obtained for the measure of assessment of the Kalman filtering algorithm's efficiency. Their respective values have been collated in Table 3 as levels of accuracy M_A, sensitivity M_S and precision M_P.

Table 3. Assessment of functional efficiency of the Kalman filtering algorithm.

Operating classes of road pavement	Measurements with using Kalman filter		
	Accuracy M_A	Sensitivity M_S	Precision M_P
Class A	94.1%	91.7%	93.9%
Class B	90.9%	84.2%	82.8%
Class C	93.6%	87.2%	85.4%
Class D	96.9%	88.9%	88.9%
All classes	93.9%	88.0%	87.7%

The Kalman filtering algorithm can be applied at the stage of road condition assessment with the accuracy of 93.9%, sensitivity of 88.0% and precision of 87.7%. The results thus obtained confirm the efficiency and precision of the solution in question, while extending the basic algorithm of pavement condition description makes it possible to eliminate the effect of isolated road defects on the overall description of the given measurement section.

5 Conclusions

When combined with the Kalman filtering algorithm, the RCT tool used in the study addressed in this paper enables road condition assessment. It provides general information concerning the assessment of pavement of the road network subject to measurements. It supports road infrastructure administration bodies in making decisions pertaining to management and maintenance of road pavements. It also allows them to identify road sections that require a more detailed assessment procedure as well as sections in need of immediate minor repair procedures or planned major repairs.

The aforementioned measurement results and road assessments obtained for the selected road sections confirm the potential of the proposed measurement method which involves application of the Kalman filtering algorithm. Implementing this solution does not require any highly advanced measuring equipment and causes no nuisance to other participants of road traffic. The assessment process is automated and may be performed by typical vehicle users in the road network.

Future research to be conducted in the scope of road condition assessment will entail taking driving styles into account, including the manner in which acceleration and braking manoeuvres are performed, for the sake of description of vehicle motion dynamics determined by the service condition of road infrastructure.

Acknowledgements. The selected research presented in this paper has been financed from the means of the National Centre for Research and Development as a part of the international project within the scope of ERA-NET Transport III Programme "Smart platform to integrate different freight transport means, manage and foster first and last mile in supply chains (S-MILE)".

References

1. Huang, S.C., Di Benedetto, H.: Advances in Asphalt Materials: Road and Pavement Construction (2015)
2. Ullidtz, P.: Pavement Analysis. Developments in Civil Engineering (1987)
3. Macioszek, E., Lach, D.: Analysis of the results of general traffic measurements in the West Pomeranian Voivodeship from 2005 to 2015. Sci. J. Sil. Univ. Technol. Ser. Transp. **97**, 93–104 (2017)
4. Macioszek, E., Lach, D.: Analysis of traffic conditions at the Brzezinska and Nowochrzanowska intersection in Myslowice (Silesian Province, Poland). Sci. J. Sil. Univ. Technol. Ser. Transp. **98**, 81–88 (2018)
5. Macioszek, E., Lach, D.: Comparative analysis of the results of general traffic measurements for the Silesian Voivodeship and Poland. Sci. J. Sil. Univ. Technol. Ser. Transp. **100**, 105–113 (2018)
6. Li, X., Goldberg, D.W.: Toward a mobile crowd sensing system for road surface assessment. Comput. Environ. Urban Syst. **69**, 51–62 (2018)
7. Staniek, M.: Stereo vision method application to road inspection. Balt. J. Road Bridg. Eng. **12**, 37–47 (2017)
8. Staniek, M., Czech, P.: Self-correcting neural network in road pavement diagnostics. Autom. Constr. **96**, 75–87 (2018)
9. Douangphachanh, V., Oneyama, H., Engineering, E.: A study on the use of smartphones for road roughness condition estimation. J. East Asia Soc. Transp. Stud. **10**, 1551–1564 (2013)
10. Du, Y., Liu, C., Wu, D., Li, S.: Application of vehicle mounted accelerometers to measure pavement roughness. Int. J. Distrib. Sens. Netw. **12**(6), 8413146 (2016)
11. Zhao, B., Nagayama, T.: IRI estimation by the frequency domain analysis of vehicle dynamic responses. Procedia Eng. **188**, 9–16 (2017)
12. Ghose, A., Biswas, P., Bhaumik, C., Sharma, M., Pal, A., Jha, A.: Road condition monitoring and alert application: Using in-vehicle smartphone as internet-connected sensor. In: 2012 IEEE International Conference on Pervasive Computing and Communications Workshops, PERCOM Workshops (2012)
13. Agostinacchio, M., Ciampa, D., Olita, S.: The vibrations induced by surface irregularities in road pavements - a Matlab® approach. Eur. Transp. Res. Rev. **6**(3), 267 (2014)
14. Zang, K., Shen, J., Huang, H., Wan, M., Shi, J.: Assessing and mapping of road surface roughness based on GPS and accelerometer sensors on bicycle-mounted smartphones. Sensors (Switzerland) **18**(3), 914 (2018)
15. Ngwangwa, H.M., Heyns, P.S., Breytenbach, H.G.A., Els, P.S.: Reconstruction of road defects and road roughness classification using Artificial Neural Networks simulation and vehicle dynamic responses: application to experimental data. J. Terramechanics **53**, 1–18 (2014)

16. Seraj, F., Van Der Zwaag, B.J., Dilo, A., Luarasi, T., Havinga, P.: Roads: a road pavement monitoring system for anomaly detection using smart phones. Lecture Notes in Computer Science (including subseries Lecture Notes in Artificial Intelligence and Lecture Notes in Bioinformatics) (2016)
17. Sierpiński, G.: Open street map as a source of information for a freight transport planning system. In: Sierpiński, G. (ed.) Advanced Solutions of Transport Systems for Growing Mobility. Advances in Intelligent Systems and Computing, vol. 631, pp. 193–202. Springer, Cham (2018)
18. Chen, G.: Introduction to random signals and applied Kalman filtering (second edition), Robert Grover Brown and Patrick Y. C. Hwang, John Wiley, New York, 1992, 512 pp., ISBN 0–47152–573–1, $62.95. Int. J. Robust Nonlinear Control **2**(3), 240–242 (2007)
19. Galińska, B.: Intelligent decision making in transport. Evaluation of Transportation Modes (Types of Vehicles) based on Multiple Criteria Methodology. In: Sierpiński, G. (red.) Integration as Solution for Advanced Smart Urban Transport Systems. Advances in Intelligent Systems and Computing, vol. 844, pp. 161–172. Springer, Cham (2019)
20. Celiński, I.: GT planner used as a tool for sustainable development of transport infrastructure. In: Suchanek, M. (ed.) New Research Trends in Transport Sustainability and Innovation: TranSopot 2017 Conference. Springer Proceedings in Business and Economics, pp. 15–27 (2018)
21. Turoń, K., Czech, P., Juzek, M.: The concept of walkable city as an alternative form of urban mobility. Sci. J. Sil. Univ. Technology. Ser. Transp. **95**, 223–230 (2017)
22. Iwan, S., Małecki, K., Korczak, J.: Impact of telematics on efficiency of urban freight transport. In: Mikulski, J. (ed.) Activities of Transport Telematics, TST 2013. Communications in Computer and Information Science, vol. 395, pp. 50–57. Springer, Berlin (2013)
23. Sierpiński, G.: Technologically advanced and responsible travel planning assisted by GT Planner. In: Macioszek, E., Sierpiński, G. (eds.) Contemporary Challenges of Transport Systems and Traffic Engineering. Lecture Notes in Network and Systems, vol. 2, pp. 65–77. Springer, Cham (2017)
24. Celiński, I.: Using GT planner to improve the functioning of public transport. Recent advances in traffic engineering for transport networks and systems. In: Macioszek, E., Sierpiński, G. (eds.) Recent Advances in Traffic Engineering for Transport Networks and Systems. Lecture Notes in Networks and Systems, vol. 21, pp. 151–160 (2018)
25. Turoń, K., Golba, D., Czech, P.: The analysis of progress CSR good practices areas in logistic companies based on reports "Responsible Business in Poland. Good Practices" in 2010–2014. Sci. J. Sil. Univ. Technology. Ser. Transp. **89**, 163–171 (2015)
26. Żak, J., Galińska, B.: Design and evaluation of global freight transportation solutions (corridors). Analysis of a Real World Case Study, Transportation Research Procedia, nr 30/2018, pp. 350–362 (2018)
27. Małecki, K., Nowosielski, A., Forczmański, P.: Multispectral data acquisition in the assessment of driver's fatigue. In: Mikulski, J. (ed.) Smart Solutions in Today's Transport, TST 2017. Communications in Computer and Information Science, vol. 715, pp. 320–332. Springer, Cham (2017)
28. Celiński, I.: Evaluation method of impact between road traffic in a traffic control area and in its surroundings. Ph.D. dissertation. Oficyna Wydawnicza Politechniki Warszawskiej, Warsaw (2018). (in polish)
29. Galińska, B.: Logistics megatrends and their influence on supply chains, business logistics in modern management. In: Proceedings of the 18th International Scientific Conference, Faculty of Economics in Osijek, pp. 583–601 (2018)
30. Golba, D., Turoń, K., Czech, P.: Diversity as an opportunity and challenge of modern organizations in TSL area. Sci. J. Sil. Univ. Technology. Ser. Transp. **90**, 63–69 (2016)

31. Sierpiński, G.: Distance and frequency of travels made with selected means of transport – a case study for the Upper Silesian conurbation (Poland). In: Sierpiński, G. (ed.) Intelligent Transport Systems and Travel Behaviour. Advances in Intelligent Systems and Computing, vol. 505, pp. 75–85. Springer, Cham (2017)
32. Staniek, M.: Repeatability of road pavement condition assessment based on three-dimensional analysis of linear accelerations of vehicles. In: IOP Conference Series: Materials Science and Engineering (2018)
33. Wan, E.A., Nelson, A.T.: Dual extended Kalman filter methods. In: Kalman Filtering and Neural Networks (2003)
34. Staniek, M.: RCT – a tool for continuous road pavement diagnostics. In: MATEC Web of Conferences (2019)
35. Regulation of General Director of National Roads and Motorways on the diagnosis of the road pavement condition and its elements, 34 (2015). (in Polish)
36. Cyganek, B., Siebert, J.P.: An Introduction to 3D Computer Vision Techniques and Algorithms (2009)

Traffic Measurements for Development a Transport Model

Marcin Jacek Kłos[(✉)], Aleksander Sobota, Renata Żochowska, and Piotr Soczówka

Faculty of Transport, Silesian University of Technology, Katowice, Poland
{marcin.j.klos,aleksander.sobota,renata.zochowska,
piotr.soczowka}@polsl.pl

Abstract. The process of building a transport model requires carrying out extensive traffic measurements and travel behaviours analyses. Part of traffic measurements is used: in the process of building a transport model, to determine structure of demand for transport and to verify the quality of the developed model. It is important to adopt an appropriate research methodology and scope of research, tailored to the needs of developing a transport model using the four-step trip-based travel demand modelling process. This article is devoted to the review of methodologies, according to which Comprehensive Traffic Survey was implemented in recent years in Poland. Case study of the construction of a transport model for the city of Bielsko-Biała was shown. The ways of representing the results of researches and measurements of road traffic have been presented.

Keywords: Traffic parameters and measurements · Transport planning · Transport model

1 Developing the Transport Model

Contemporary management of transport systems requires development of a road traffic model in order to determine the current state of the transport network. Models of the current state of the network allow to provide data to facilitate decision-making regarding changes of: traffic organization, geometry of the intersections and roads and control systems [1]. The indicated data are measures of effectiveness assessment, e.g. transport performance, travel time, capacity utilization rate. The model also allows to perform: assessment of transport system functioning in a given area, analysis and forecast of traffic volumes, determination of the size of passenger flows and effective planning and management of the transport system. Three types of transport model exist [2]. The main criterion of division is the area of coverage. Hence, the following types of models may be distinguished:

- macroscopic models – based on the deterministic relationships of the flow, speed, and density of the traffic stream,
- microscopic models – that take into account the characteristics of every individual object of the population for determining the travel demand of the region,

- mesoscopic models – that combine the properties of both microscopic and macroscopic models.

This article is related to the analysis of traffic measurements necessary for develop a macroscopic model. Analysis of the measurements for the models of different area in Poland was carried out. Detailed case study for model developed for the city of Bielsko-Biała was presented. The indicated macroscopic model includes defined parameters for vehicle groups, it does not describe individual vehicles.

Figure 1 presents the method of developing a transport model, so-called four-step trip based travel demand modelling process.

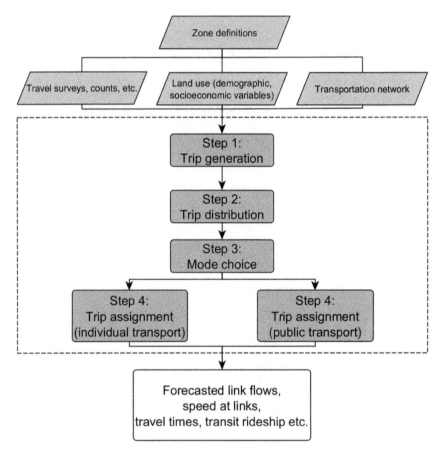

Fig. 1. Scheme for four-step model - transport demand modelling process (source: [3])

In order to perform four-step model processes, input data connected with traffic measurements, surveys, transport network and others are determined. The four-step travel demand modelling consists of four main stages [4]:

- trip generation model – determining the number of trips generated and absorbed depending on land use, demographic and economy activity data in individual TAZ (Traffic Analysis Zone) within the individual travel purposes: (HBW) home based work, (HBNW) home based non-work (e.g. school, shopping, recreation, other) and non-home based trips,
- trip distribution model – development of a spatial trip distribution on the basis of a travel demand matrix (description about zone of start and end of trip),
- modal split model – describes the criteria for the selection of individual available means of transport for the travel; usually the choice is made between five ways of selecting transport means: walking, car as a driver, car as a passenger, public transport and bicycle,
- trip assignment model – defines travel routes within individual transport systems, and consequently determines the intensity of traffic and passenger flows.

Table 1 presents a list of traffic flows surveys and measurements carried out to develop a travel demand model based on a literature review [3–7]. The types of necessary data and measurements were divided into eight main categories related to: network, socio-economic data, trip generation, trip distribution, time of day of travel, mode choice, transit assignment and measures of model fitting.

Table 1. Lists of traffic flows surveys and measurements to develop travel demand model.

Component	Traffic flows data, surveys and measurements to develop a road traffic model
Network	– node and link identifiers and functional classification – distance and speed on links – status of links – one-way or two-way status – network topology – zone size, type and structure compliance – network density and connectivity – transit run times
Socio-economic data	– location of special generators – population by geographic area – distribution of households and jobs (by employment and geographical location) – education – car ownership
Trip generation	– factors affecting the generation (absorption) of travel – trip rates (produced and attracted) for each zone – number of trips generated and absorbed in each zone – density of land use
Trip distribution	– trip length frequency distribution – work-based trips by zone – zone-to-zone flow/desire lines – origin-destination surveys by trip purpose

(*continued*)

Table 1. (*continued*)

Component	Traffic flows data, surveys and measurements to develop a road traffic model
Time of day of travel	– time of day versus volume peaking – cordon counts – speed by time of days
Mode choice	– mode shares – parameters of level of services
Transit assignment	– major station boarding – bus line – parking infrastructure – transits – load factors
Model calibration	– assigned versus observed number of vehicles by screenline or cutline points – assigned versus observed speeds of vehicles – assigned versus observed number of vehicles by direction and time of day – assigned versus observed number of vehicles by vehicle class – cordon line volumes

The process of building a transport model requires a series of researches, which should be connected to the features of Comprehensive Traffic Survey. The results are used to build the travel demand model (e.g. household surveys) or to map characteristics of demand of the transport system (e.g. research of the link impedance function). Others are used to check the quality of the constructed model, e.g. occupancy rate in public transport vehicles, traffic volume measurements at screenline points and node points of the transport network. It is therefore important to adopt an appropriate methodology for the Comprehensive Traffic Survey.

The paper is organized as follows, Sect. 2 provides the scope of conducted research and traffic measurements for the construction of transport models in Poland. The next section contains a case study in which measurement points and visualization of data carried out to develop a transport model for the city of Bielsko-Biała were presented. Section 4 contains the results of determining of the peak hour in the travel demand model. The next section presents the results of the percentage share of the structure of vehicle groups in the indicated case study. Sections 6 of the paper contains conclusions and propositions for future work.

2 Research for Building and Development of the Transport Model in Poland

In the process of building a transport model for a city, a number of different research are carried out that allow map of the current network of the analysed area. This article describes the method of obtaining and analysing the results of road traffic research for transport model. An analysis of the scope of research and traffic measurements for the building of transport models in Poland was carried out. Table 2 presents areas analysed with a demographic description and area coverage.

Table 2. Areas subjected to the analysis of performed research and traffic measurements for the construction of transport models in Poland, population and area coverage.

Area	Population [-]	Area coverage [km^2]
Bielsko-Biała [8]	170.500	124.000
Katowice [9]	302.400	164.000
Subregion of the Central Silesian Voivodeship [10]	2.784.951	5.577
Warszawa [11]	1.764.615	517.000

Table 3 presents the characteristics of the conducted research for the analysed areas, which were made for the purpose of developing a road traffic model.

Table 3. Characteristics of the conducted research for the analysed areas, which allow to develop a transport model.

City	Characteristics of conducted research and measurements of road traffic
Bielsko Biała [8]	– surveys in households – 5,300 [-] – traffic volume for cordon – 16 points – traffic volume for screenline points – 19 points – traffic volume in nodes – 79 intersections – traffic generation by shopping centres and large-surface parking lots – 32 locations – occupancy rate in public transport vehicles (each course on the entire route) – occupancy rate at railway and bus stations – 20 locations – surveys of public transport passengers on the cordon of cities (all entry and exit courses from the city)
Katowice [9]	– surveys in households – 5,000 [-] – traffic volume in nodes – 70 intersection – occupancy rate in public transport vehicles – measurement of transit relations – surveys of public transport passengers on the cordon of cities
Subregion of the Central Silesian Voivodeship [10]	– surveys in households – 48,367 [-] – surveys of public transport passengers on the cordon of cities – surveys in shopping centres and large-surface parking lots – 500 surveys per location – surveys of public transport passengers on the cordon of cities – 50 surveys per location – surveys of freight traffic – 2,500 surveys – occupancy rate in public transport vehicles – traffic volume for screenline points – 74 points – traffic volume for cordon – 83 points – traffic volume in nodes and links – link impedance function

(*continued*)

Table 3. (*continued*)

City	Characteristics of conducted research and measurements of road traffic
Warszawa [11]	– surveys in households – 17,000 [-] – surveys of public transport passengers on the cordon of cities – 1,000 surveys – surveys of freight traffic – 1,500 surveys – traffic volume for cordon in Warszawa – 55 points – traffic volume for cordon in agglomeration – 67 points – traffic volume for screenline – 30 points – traffic volume in nodes and links – 64 points – occupancy rate in public transport vehicles

The presented types of road traffic research and measurements carried out in Polish cities (Table 3) show the necessary data for the development of transport model using the four-step model process (Table 1). Number of necessary measurements connected with surveys and number of points is different and depends on transport model area coverage. However detail level of the transport model is dependent on the range of the research and measurements.

3 Case Study – Measurement Points and Data Visualization

The case study includes detailed data on completed researches and traffic measurements aimed at developing the transport model for the city of Bielsko-Biała. The presented data were obtained during the performance of a scientific and research project involving the transport model for the city of Bielsko-Biała [8].

The methods of acquiring traffic data were different depending on the selected object. The following measuring tools were used: stationary cameras, cameras mounted at intersections, and additionally trained employees performing measurements using special measurement sheets.

In order to obtain traffic results, measurement points and time of research were determined, data are presented in Table 4. The number and place of conducted measurements is important for the quality of the developed transport model.

Table 4. The number of individual measurement points determined by objects of the research.

Research objects	Number of measurement points [-]	Time period of research [hh: mm]
Intersections	79	06:00–18:00
Cordon points	16	00:00–24:00
Screenline points	19	06:00–18:00

Measurements conducted at intersections included important traffic-related points located in the city. A large number of measurement points allows for accurate mapping of the current state of traffic in the city [12]. Dense number of points also allows to track the distribution of traffic flows.

Standard intersections with three or four inlets and developed road junctions have been analysed. In order to accurately and easily interpret the acquired data for intersections, it is good practice to develop flow maps, which describe the intensity of road traffic occurring at individual inlets along with the division into directions of the links [13]. Figure 2 shows a flow map for a road node located on the S52 road near the intersection with Bielska Street.

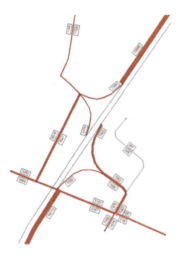

Fig. 2. An example of a road node flow map (for the entire measurement period) showing the traffic intensity distribution with a division into directions of the links

Figure 3 shows the method of developing a flow map, which describes a standard intersection, while Fig. 2 illustrates the presentation of a flow map for an extended road node. Both Figs. 2 and 3 were made using the VISUM software [14].

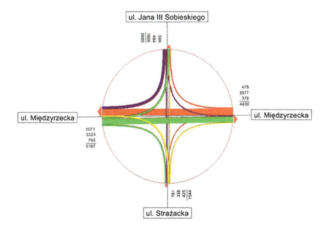

Fig. 3. An example of flow map for an intersection presenting the traffic intensity (for the entire measurement period) with a division into directions of the links

The cordon measurements are carried out in order to obtain data on the transit traffic of the city. During the measurements, in addition to basic parameters such as type and number of vehicles, the vehicle plate number must be identified. Basing on plate number data, it is possible to make and illustrate (Fig. 4) transit traffic.

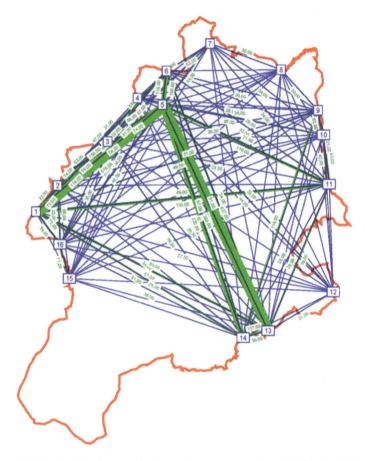

Fig. 4. Road traffic structure for the entire measurement period for transit trips

A few transit traffic flows are more heavily loaded than others. The highlighted points 1, 5, 13, 14 are correspond with the following directions of movement of vehicles:

- 1: cities: Skoczów, Cieszyn, Ustroń, Wisła,
- 5: cities: Pszczyna, Tychy, Katowice, Warszawa,
- 13: cities: Żywiec, Korbielów, Zwardoń,
- 14: cities: Wisła, Szczyrk.

Figure 5 shows intra-city traffic flows for the analysed period of time based on the results of screenline points measurements.

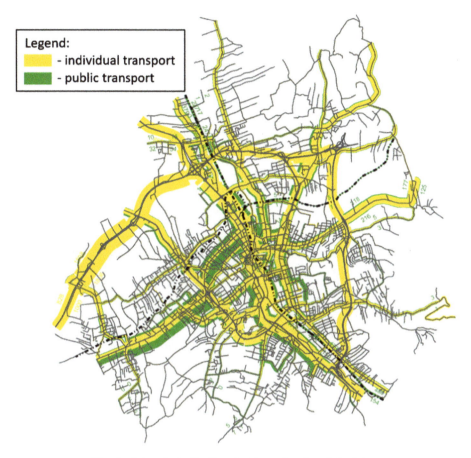

Fig. 5. Intra-city traffic flow for the analysed period of time

Screenline points research shows the intensity of traffic in inner cities. The points designated for these measurements present the intersections of the road with the railway line.

4 Case Study: Determination of Peak Hours for the City

One of the main goals of conducting road traffic research for transport model is to determine the peak hour for the entire analysed area. The identification of peak hours and detailed traffic research allows to perform the distribution of intensity of traffic flows and find the location of the nods or links in the road network with the highest load.

Analysis of the results of road traffic research starts with the peak hours determination for the individual measurement point. Afterwards peak hours are aggregated into sets of results for type of research. The final stage is to determine the global peak hour for the entire analysed area. In reference to the methodology of determining the peak hour, two time periods are sought for the morning and afternoon rush [15]. Before calculations of peak hours, the heterogeneous traffic flow is converted into homogeneous one using Passenger Car Units (PCUs), to account for the differential impact of different types of vehicles on the traffic volume. Table 5 presents a summary of identified peak hours divided into types of research.

Table 5. Identified peak hours divided into objects of research.

Research objects	Morning peak hour	Afternoon peak hour
Intersections	07:15–08:15	15:15–16:15
Cordon points	06:45–07:45	13:30–14:30
Screenline points	07:30–08:30	15:30–16:30

Analysis of the results shows an earlier occurrence of the peak hour for cordon points in relation to the screenline points and a summary list of all research. Calculation of the peak hour for cordon points is related to traffic on the inlets (transit, destination) and outlets from the city (source and transit). Peak hours were also determined during the survey: morning peak hour 07:30–08:30 and afternoon peak hour 14:30–15:30 [5]. The time period obtained in the survey is similar to the identified peak hours in research objects. Figures 6 and 7 shows the distribution of traffic intensity for screenline and cordons points.

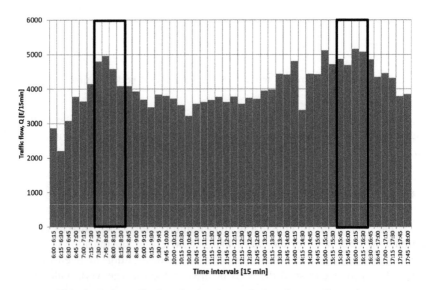

Fig. 6. Daily traffic flow in time intervals for all screenline points

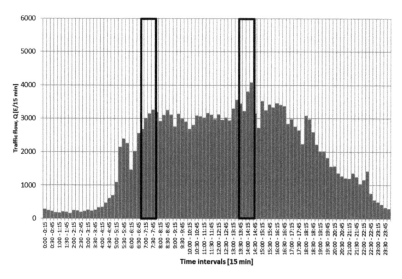

Fig. 7. Daily traffic flow in the time intervals for all cordon points

Black frames depicted the occurrence of peak hours (morning and afternoon) in the figures. For intra-city the sum of traffic flow is bigger than in city cordons. The difference is 8,234 equivalent vehicles.

Figure 8 shows the distribution of traffic flow in interval times for the entire city of Bielsko-Biała with the indicated peak hours.

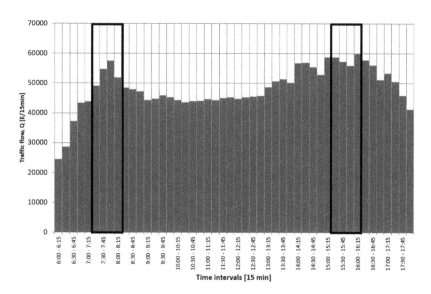

Fig. 8. Peak hours for city Bielsko-Biała with all research results

The highest volume of traffic flow during the morning peak hour is 181,232 equivalent vehicles per hour. The highest intensity in the afternoon peak hours is 198,223 equivalent vehicles per hour.

5 Case Study: Percentage Share of the Structure of Vehicle Groups

In order to compare the traffic characteristics between the cordon and screenline points, the analysis of the percentage share of the structure of vehicles groups was carried out. Analysis of the structure allows identification of the specificity of traffic flow in the analysed area. Input this results to the transport model allows for effective planning of traffic and investments related to road infrastructure.

Figures 9 and 10 show the percentage share of the structure of vehicles groups for all screenline and cordon points.

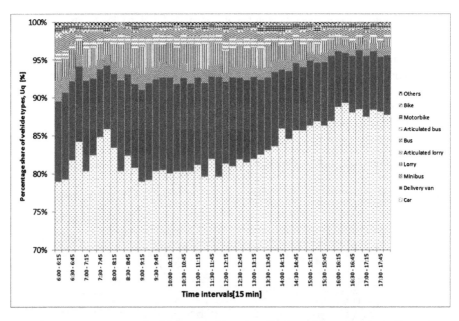

Fig. 9. Percentage share of vehicle types in total traffic flow observed at screenline points

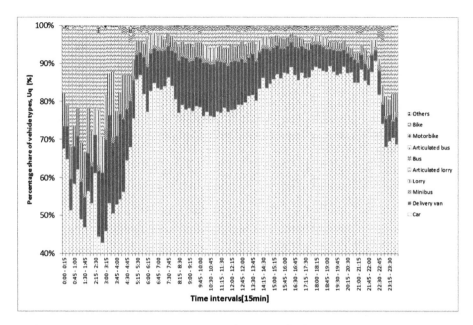

Fig. 10. Percentage share of vehicle types in total traffic flow observed at cordon points

Results for percentage share of vehicles groups show a higher shares of traffic for minibuses (0.7% difference), buses (difference by 0.3%) and motorbikes (difference by 0.3%) was observed for the screenline points. For the cordon points the increased traffic of heavy goods vehicles was identified (difference of 3.3%). The flow of passenger cars for two types of research is similar both for cordon is 82.3% and for screenline points 83.7%.

6 Conclusion

Transport model is an extremely important tool, which allows to efficiently manage and enables planning of changes in road infrastructure in the city. The implementation of such model is related to the conduct of a number of research. The part of analyses described in the paper concerns on the road traffic measurements.

The number and scope of measurements for a given area determine the level of detail of the results obtained. A dense network of measurement points enables obtaining reliable results efficient for the road network model determining.

During the research, an important element is the method of presenting and visualizing the results. The described methods, such as: flow map, road traffic structure, distribution of traffic flow allow easy interpretation of complex data.

The identification of the peak hour for the city is a basic aspect, which allows the distribution of flow streams to the road network at the most-charged moment of the day. Visualization of traffic streams allows to attempt to take measures for improve traffic in locations with the highest level of congestion. During the identification of peak hours it is also necessary to determine a distribution of the traffic intensity. The illustrated flow allows to check whether the peak hour significantly differs in the volume of traffic received from the rest of the distribution.

Presentation of the percentage share of vehicle types allows to analyse changes in the traffic characteristics between the examined objects.

Future research on the development of a transport model will include the analysis of the results obtained in surveys. Another aspect considered will be studies related to the distribution of traffic flows depending on the selected prognostic variants with developed transport model.

References

1. Kłos, M.: Identifications peak hours on intersections set in Bielsko-Biała City. Sci. J. Sil. Univ. Technol. **90**, 113–121 (2016)
2. Li, L., Chen, X.M.: Vehicle headway modeling and its inferences in macroscopic/microscopic traffic flow theory: a survey. Transp. Res. Part C Emerg. Technol. **76**, 170–188 (2017)
3. Wolshon, B., Pande, A.: Traffic Engineering Handbook. Wiley, Hoboken (2016)
4. Abadi, A., Rajabioun, T., Ioannou, P.A.: Traffic flow prediction for road transport networks with limited traffic data. IEEE Trans. Intell. Transp. Syst. **16**(2), 653–662 (2015)
5. Meyer, M.D.: Transport Planning Handbook. Wiley, Hoboken (2016)
6. Black, J.: Urban Transport Planning: Theory and Practice. Routledge, Abingdon (2018)
7. Żochowska, R., Karoń, G., Janecki, R., Sobota, A.: Selected aspects of the methodology of traffic flows surveys and measurements on an urban agglomeration scale with regard to ITS projects. In: Macioszek, E., Sierpiński, G. (eds.) Recent Advances in Traffic Engineering for Transport Networks and Systems. Lecture Notes in Networks and Systems, vol. 21, pp. 37–49. Springer, Cham (2018)
8. Sobota, A., Janecki, R., Karoń, G., Żochowska, R., et al.: Zintegrowany System Zarządzania Transportem na obszarze miasta Bielska-Białej, etap I-wykonanie Modelu Ruchu dla miasta Bielsko-Biała. Praca NB-148/RT5/2014, Wydział Transportu Politechniki Śląskiej, Katowice (2015). (in Polish)
9. Franek, Ł., Struska, O., Szpórnóg, M., et al.: Wieloletni plan rozwoju zintegrowanego systemu transportowego miasta Katowice (2016). (in Polish)
10. Thiem, J., Thiem, J., Maćkowiak, A. et al.: Studium Transportowe Subregionu Centralnego Województwa Śląskiego (2018). (in Polish)
11. Żak, J., Jachimowski, R., Kłodawski, M., Lewczuk, K. et al.: Warszawskie Badanie Ruchu 2015 wraz z opracowaniem modelu ruchu (2015). (in Polish)
12. Te Brömmelstroet, M., et al.: Experiences with transportation models: an international survey of planning practices. Transp. Policy **58**, 10–18 (2017)
13. Arliansyah, J., Prasetyo, M.R., Kurnia, A.Y.: Planning of city transportation infrastructure based on macro simulation model. Int. J. Adv. Sci. Eng. Inf. Technol. **7**(4), 1262–1267 (2017)
14. Ptv, A.G.: PTV Visum Manual. Karlsruhe, Germany (2015)
15. Gaca, S., Suchorzewski, W., Tracz, M.: Inżynieria ruchu drogowego. Teoria i praktyka. Wydawnictwo Komunikacji i Łączności, Warszawa (2008). (in Polish)

Life Cycle Sustainability Assessment of Sport Utility Vehicles: The Case for Qatar

Nour N. M. Aboushaqrah[1], Nuri Cihat Onat[1(✉)], Murat Kucukvar[2], and Rateb Jabbar[1]

[1] Qatar Transportation and Traffic Safety Center, College of Engineering, Qatar University, Doha, Qatar
onat@qu.edu.qa
[2] Mechanical and Industrial Engineering, College of Engineering, Qatar University, Doha, Qatar

Abstract. Electric vehicle technologies are attractive alternatives to traditional vehicles towards achieving sustainable transportation. The adoption of these technologies has a great potential in reducing road transportation externalities. As Qatar aims to achieve 10% electric vehicles by 2030, this research reveals the macro-level environmental, social, and economic impacts and benefits of electric vehicles in Qatar. The studied vehicle technologies are, gasoline vehicle (ICV), hybrid electric vehicle (HEV), plug-in hybrid electric vehicle (PHEV), and battery electric vehicle (BEV). In this regard, we quantified 9 macro level indicators using Multi regional input-output (MRIO)-based life cycle sustainability assessment (LCSA) framework and compared the vehicles accordingly. The results show that, electric vehicles are better options in terms of Global Warming Potential (GWP), Particulate Matter Formation (PMF), and Photochemical Ozone Formation (POF) impacts. In addition to that, the results demonstrated that electric vehicles are more cost effectives than the traditional ones, while they are worse than traditional vehicles in terms of employment, operating surplus, and Gross Domestic Product (GDP) impacts.

Keywords: Life-cycle sustainability assessment · Decision-support · Multi-regional input-output analysis · Sustainable transportation · Electric vehicles

1 Introduction

The global climate change and air pollution, besides fossil fuel dependency are some of the leading challenges towards developing sustainable transportation [1]. In the context of these challenges, there is an urge to search for alternative vehicle technologies to conventional gasoline vehicles in order to mitigate the environmental, economic, and social impacts towards achieving sustainable mobility. In fact, electric vehicle technologies such as battery electric vehicles (BEV), plug-in-hybrid electric vehicles (PHEV), and hybrid electric vehicles (HEV) are considered attractive alternatives due to their great potential of reducing air pollution, global climate change, associated health impacts on residents, and fossil fuel consumption [2]. Many countries around the world are increasingly moving toward the adoption of electric vehicle technologies, and similarly, Qatar doesn't want to miss out this trend and therefore Qatar has set a goal of 10% electric vehicle sales by 2030 [3]. In contrast, there are some certain barriers against the adoption of these technologies, including the lack of charging infrastructure, charging time, high initial price of BEVs, range anxiety, operational issues and uncertainties associated with the possible benefits of these technologies [4].

Qatar's 2030 national vision aims to create a balance between economic and social growth with environmental development, which are closely aligned to the three pillars of sustainability [5]. In fact, Qatar is the world's largest CO_2 emitter per capita as per the Intergovernmental Panel on Climate Change (IPCC) reports [6]. Besides, Doha is ranked the 12th among the most polluted cities in the world, according to the World Health Organization, in terms of the annual mean concentration of particulate matter formation (PM2.5 and PM10) which poses serious risks to human health [7]. Additionally, transportation sector is one of the major sectors contributing carbon emissions and air pollution at a global scale, the same in Qatar, as a result of the very high automobile dependency, and the lack of means of transportation and alternative vehicle technologies [8]. Indeed, while Qatar needs to pursue improving the access to goods and services to foster the economic and social growth, simultaneously, Qatar should mitigate the impacts of transportation on environment, economy, and society, the three pillars of sustainable development, towards a more sustainable and efficient mobility. To this end, this research is an important attempt to minimize the transportation externalities of road transport in Qatar, by evaluating the potential environmental, economic, and social impacts and benefits.

Life cycle assessment (LCA) is a widely known method that assesses the environmental impacts related to a product's/service's life cycle from the extraction of the raw materials to final disposal, however, this method performs the environmental assessment in an isolated way with no consideration of economic and social impacts [9]. Therefore, due to major shortcomings of LCA, there has been a transition from LCA to Life Cycle Sustainability Assessment (LCSA), where environmental life cycle assessment (LCA), economic life cycle costing (LCC) and social life cycle assessment (S-LCA) methods are used together for an integrated sustainability assessment [9]. Life cycle sustainability assessment (LCSA) aims to assess the environmental, economic and social dimensions of sustainability which are also called the triple bottom lines (termed as TBL) of sustainable development [10].

LCSA is still a relatively new framework and needs further developments especially in terms of applications. A minority LCSA studies in the literature have used LCSA in real case studies for product LCSA [11]. Practical applications of LCSA requires the integration of various system-based tools such as, Multi regional input-output (MRIO) approach, system dynamics modeling, and multi criteria decision making tool [11]. Multi Region Input–Output (MRIO) is an advanced version of single region IO modeling approach, and the integration of MRIO molding with LCSA framework allows to capture the global supply chain-related impacts [11]. On the other hand, the integration of MRIO with LCSA is very limited in the literature [11]. Therefore, the use of MRIO within LCSA framework is achieved in this study to holistically analyze the sustainability impacts of alternative vehicle technologies.

According to the literature, many studies have analyzed the sustainability assessment of electric vehicle technologies. For example, Onat et al. [12] have developed a hybrid LCSA model to evaluate the macro-level environmental, economic and social impacts of alternative passenger vehicle technologies in the U.S. In several studies, the LCSA is applied for sustainable transportation, and used a holistic hybrid LCSA method to quantify the macro-level environmental, economic and social impacts of alternative passenger vehicles in the U.S. [13].

To this end, this paper uses a novel Multi-regional input-output based life-cycle sustainability assessment framework (LCSA) to quantify the macro-level economic, social, and environmental impacts associated with the entire global supply chain for the alternative vehicle technologies in Qatar.

2 Methodology

In this research, direct impacts (tailpipe emissions) and upstream impacts related to inside Qatar fuel supply, inside Qatar sectors, and outside Qatar sectors, are calculated for alternative vehicle options using multi-regional input-output (MRIO) based life cycle sustainability assessment (LCSA). 9 macro-level indicators representing the environmental, economic, and social impacts are quantified. Sport utility vehicle (SUV) body style is considered in this study representing four different vehicle technologies, internal combustion vehicles (ICV), hybrid electric vehicles (HEV), plug-in-hybrid electric vehicles (PHEV), and full battery electric vehicles (BEV). The considered vehicles are compared based upon the quantified indicators. The analysis covers the vehicle's operation phase as it is the most dominant phase in terms of the environmental impacts, while the other life cycle phases are excluded from the scope of this paper. In operation phase analysis, two main sub-phases, Well-to-Tank (WTT) and Tank-to-Well (TTW) are studied. WTT represents the upstream impacts related to inside Qatar fuel supply (gasoline/electricity supply chain), inside Qatar sectors, and outside Qatar sectors. While, TTW refers to the direct impacts (tailpipe emissions) during vehicle travel. The functional unit of this analysis is 1 km of vehicle travel.

2.1 Well-to-Tank Operation Sub-phase

WTT calculations are made for both gasoline and electricity supply, as they are the main components for the studied vehicles. Gasoline is consumed by ICVs, HEVs and PHEVs and the upstream impacts associated with the gasoline supply chain is calculated for each vehicle type using the upstream impact factors associated with the production and combustion of 1 L of gasoline. These impact factors are obtained from EXIDBASE emission database. The gasoline amount required to travel 1 km (fuel economy) for each vehicle is also determined for the upstream impact calculation purpose. The fuel economy for the studied ICV and HEV are 14.5 and 7.84 L per 100 km (L/100 km) respectively. Whereas the fuel economy for PHEV is 9.8 L/100 km in gasoline mode and 37.4 kWh/100 km in electric mode, however the fuel economy for EV is 22.5 kWh/100 km. The total upstream impacts (WTT) related to gasoline supply is calculated by multiplying the fuel economy of each vehicle type with the associated upstream impact factors resulting from the inside Qatar gasoline supply, inside Qatar sectors, and outside Qatar sectors. The calculation of WTT impact per km of travel is shown in Eq. 1.

$$(SI)_i = (FE)_j * [(IF_{\text{fuel supply}})_i + (IF_{\text{inside Qatar sectors}})_i + (IF_{\text{outside Qatar sectors}})_i] \quad (1)$$

Where, SI: represents sustainability impact, i: represents the sustainability impact category, j: represents vehicle type, FE: represents the fuel efficiency of a vehicle, IF: represents Impact factors obtained from the MRIO model.

Electric power supply is the second main component of WTT analysis and is used to operate EVs and PHEVs. The upstream impact factors to generate per kWh of electricity from Natural gas are taken from EXIDBASE database. The WTT impacts related to electricity supply are calculated by multiplying the fuel economy of each vehicle type with the associated upstream impact factor. On the other hand, the calculation of the WTT impacts associated with PHEVs is performed differently as they operate on both gasoline and electric power mode. The portion that PHEV use to operate in electric mode is estimated by the utility factor (UF), that is the percentage of the distance traveled in electric mode. The utility factor of PHEV-22 km in Qatar is estimated to be 35%.

2.2 Tank-to-Wheel Operation Sub-phase

TTW impacts are directly associated with the amount of gasoline/electricity consumed during the vehicle operation. TTW impacts are calculated through multiplying the fuel efficiency of each vehicle type with the associated tailpipe emission factors gathered from GREET model. On the other hand, it is important to note that BEVs and electric mode portion of PHEVs do not have any tailpipe emissions and therefore have zero TTW impacts.

2.3 Life Cycle Cost Analysis

Life-cycle costing (LCC) refers to the total cost of ownership over the life of a project, product or service. This analysis includes various cost elements such as, initial costs, annual fuel costs, maintenance costs, insurance costs, and salvage values of the studied alternatives vehicles. Economic factors such as interest rate, inflation rate, inflation adjusted interest rate, and vehicle depreciation are also considered in this study. In the analysis, the average annual travel distance is estimated to be 22,000 km, collected from FAHES, a local semi-government agency gathering meter readings in Qatar. The average gasoline and electricity prices are gathered from local agencies. The cost elements considered, and the key assumptions made for performing this analysis are presented in Table 1.

Table 1. Summary of the cost elements considered and the key assumptions made for the life cycle cost analysis.

Life cycle costs	ICV	HEV	PHEV	(EV)
Initial price (QR)	QAR 259,000	QAR 199,344.75	QAR 232,145.63	QAR 289,499.25
Fuel usage (L/100 km)	14.5 Lt	7.84 Lt	37.4 kwh electricity 9.8 L gasoline	22.5 kwh
Average price per liter (QR/L)	2	2	2 QR/L 0.085 QR/kwh	0.085 QR/kwh
Annual fuel costs (QR/Year)	6376.82 QAR/year	3448.28 QAR/year	Electric mode: 699.33 QR/year Gasoline mode: 4313.72 QR/year Total Cost: 0.35(699.33) + 0.65 (4313.72) = 3048.68 QAR/Year	421.17 QAR/year
Average maintenance costs (QR/Year)	283.1 QAR	374.43 QAR	728.22 QAR	748.85 QAR
Insurance cost per year (QR/Year)	4% initial cost	4% initial cost	4% initial cost	4% initial cost
Salvage value (QR)	0.25 of initial cost	0.25 of initial cost	0.25 of initial cost	0.25 of initial cost
Interest rate	4.78%	4.78%	4.78%	4.78%
Inflation rate	2.93%	2.93%	2.93%	2.93%

The analysis steps are as follows: First, the vehicle life expectancy, cost element costs in Qatari Riyal (QAR), and economic factors for each vehicle alternative are estimated. Next, present worth analysis is used to calculate the present value for each cost element for every year. Finally, the LCC is calculated by adding the costs of each cost element at the end of each year. In the end, the findings of each vehicle are analyzed and compared and the lowest LCC vehicle alternative is selected. This analysis is conducted to select the most cost-effective option (the lowest LCC) in Qatari Riyal per kilometer (QAR/Km) over the vehicle's lifespan.

3 Results and Discussions

3.1 Environmental Impacts

The environmental impacts of each vehicle technology are presented in Fig. 1. According to the results, the tailpipe (direct) impact is the highest contributor to GWP, PMF, and POF for all vehicle types except for BEV, as BEV does not produce any tailpipe impact. The highest amount of Global Warming Potential (GWP), Particulate Matter Formation (PMF), and Photochemical Ozone Formation (POF) associated with TTW comes from ICVs, while the least amount produced in this stage comes from PHEVs. However, inside Qatar fuel supply is the highest contributor to GWP, PMF, and POF for BEV, as opposed to HEV which has the least impact generated from this stage. Very few GWP, PMF, and POF impacts are generated from inside and outside Qatar sectors for all vehicle types. Overall, ICV has the highest contribution to GWP, PMF, and POF impact per km, followed by PHEV, followed by HEV, and finally followed by BEV.

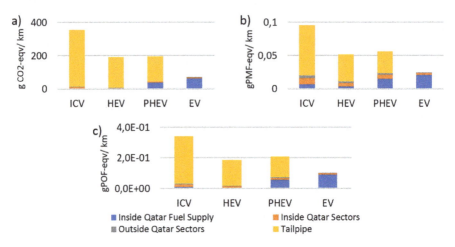

Fig. 1. Environmental impacts of alternative vehicle technologies: (a) GWP (gCO2 eqv. per km); (b) PMF (gPMF-eqv. per km); (c) POF (gPOF-eqv. per km)

3.2 Social Impacts

Figure 2 presents the social impacts for each vehicle technology. The results show that, ICV, HEV, and PHEV have the highest Human health impact associated with (TTW) stage, while no vehicle has total tax and employment impacts generated from this stage. However, inside Qatar fuel supply dominates the total tax and employment impacts for all vehicle options, where ICV accounts for the highest impact, and BEV has the least. Moreover, inside Qatar fuel supply is the highest contributor to human health impact for BEVs. In addition, outside Qatar sectors is an important contributor to employment impact category, as opposed to inside Qatar sectors. As illustrated, all vehicle types have substantially low total tax and human health impacts inside and outside Qatar sectors. Overall, BEV has the least total tax impact, followed by PHEV as the second least total tax impact, followed by HEV, and finally followed by ICV. Besides, BEV has the least contribution to employment, followed by HEV, followed by PHEV, followed by ICV. While, BEV has the least human health impact, followed by HEV, followed by PHEV, and finally followed by ICV.

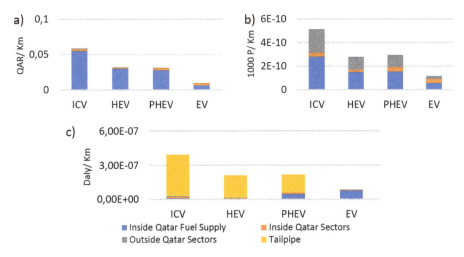

Fig. 2. Social impacts of alternative vehicle technologies: (a) Total Tax (QAR per km); (b) Employment (1000 P per km); (c) Human Health (Daly per km).

3.3 Economic Impacts

Figure 3 shows the economic impacts. No vehicle has operating surplus and Gross Domestic Product (GDP) impacts generated from (TTW) stage. The fuel supply inside Qatar dominates the total impacts of operating surplus and GDP for ICVs, HEVs, and PHEVs, where ICV has the highest operating surplus and GDP impacts, and BEV

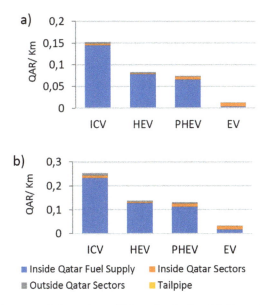

Fig. 3. Economic impacts of alternative vehicle technologies: (a) Operating surplus (QAR per km); (b) GDP (QAR per km).

generates the least. However, BEV has the highest contribution to operating surplus and GDP impacts for inside Qatar sectors. The results show that, very low operating surplus and GDP impacts are generated from inside and outside Qatar sectors for all vehicle types.

3.4 Life Cycle Cost

The comparison between the studied alternative technologies in terms of total vehicle ownership costs is shown in Fig. 4. According to the results, the initial cost of BEV is the highest (1.097 QAR/Km), while HEV has the least initial cost (0.755 QAR/Km). The fuel costs vary significantly among the alternatives and ranges between 0.312 QAR/km (ICV) and 0.023 QAR/km (BEV), while salvage values vary insignificantly between 0.222 (BEV) and 0.152 (HEV). The results show that, BEV has the highest maintenance (0.037) and insurance costs (0.367), while ICV has the lowest maintenance cost (0.014) and HEV has the lowest insurance cost (0.253). Overall, HEV has the least cost followed by PHEV followed by BEV and finally followed by ICV.

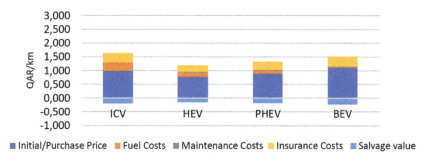

Fig. 4. Life cycle cost analysis (QAR/km)

4 Conclusion and Future Work

In this research, 9 macro-level indicators representing environmental, economic, and social impacts are quantified using a multi-regional input-output (MRIO) based life cycle sustainability assessment (LCSA) framework. Four types of Sport Utility Vehicles (SUVs) were analyzed and compared according to the quantified indicators during the operation phase. According to the results, electric vehicle technologies are superior to conventional vehicles in terms of Global Warming Potential (GWP), Particulate Matter Formation (PMF), and Photochemical Ozone Formation (POF) impacts. Besides, electric vehicles are more economical options in comparison with conventional vehicles. On the other hand, the results showed that electric vehicles are worse than gasoline vehicles in terms of employment, operating surplus, and Gross Domestic Product (GDP) impacts. The developed MRIO based LCSA model can be further improved by inclusion of system dynamics modeling to reveal the interconnections and the dynamic relationships between the indicators, and inclusion of Multi criteria decision making tools as they are powerful for selecting the optimum alternative among many alternatives with multiple objectives.

Acknowledgment. This paper is an output of a project supported by Qatar University, grant number QUSD-CENG-2018\2019-2.

References

1. Yuksel, T., Tamayao, M.A.M., Hendrickson, C., Azevedo, I.M., Michalek, J.J.: Effect of regional grid mix, driving patterns and climate on the comparative carbon footprint of gasoline and plug-in electric vehicles in the United States. Environ. Res. Lett. **11**(4), 044007 (2016)
2. Bicer, Y., Dincer, I.: Comparative life cycle assessment of hydrogen, methanol and electric vehicles from well to wheel. Int. J. Hydrogen Energy **42**, 3767–3777 (2017)
3. Qatar General Electricity and Water Corporation. MoU signed to launch Green Car Initiative (2017)
4. Qatar Green Leaders. The Next Big Obstacle for Electric Vehicles? Charging Infrastructure (2017). http://www.qatargreenleaders.com/news/sustainability-news/1438-the-next-big-obstacle-for-electric-vehicles-charging-infrastructure. Accessed 12 Dec 2017
5. The Ministry of Development Planning and Statistics. Qatar National Vision 2030 (2018)
6. Our World in Data CO_2 and other Greenhouse Gas Emissions (2019)
7. JustHere Doha. Doha 12th most polluted city in the world; possible causes are higher air traffic and growing population (2014). http://www.justhere.qa/2014/05/doha-among-worlds-polluted-cities-higher-air-traffic-population-levels-may-causal-agents/
8. Qatar Rail. Qatar Rail Annual Report 2015–2016 (2016)
9. Guinée, J.: Life cycle sustainability assessment: what is it and what are its challenges? In: Taking Stock of Industrial Ecology, pp. 45–68. Springer, Cham (2016)
10. Guinee, J.B., Heijungs, R., Huppes, G., Zamagni, A., Masoni, P., Buonamici, R., Ekvall, T., Rydberg, T.: Life cycle assessment: past, present, and future †. Environ. Sci. Technol. **45**, 90–96 (2011)
11. Onat, N.C., Kucukvar, M., Halog, A., Cloutier, S.: Systems thinking for life cycle sustainability assessment: a review of recent developments, applications, and future perspectives. Sustainability **9**, 706 (2017)
12. Onat, N.C., Kucukvar, M., Tatari, O.: Towards life cycle sustainability assessment of alternative passenger vehicles. Sustainability **6**, 9305–9342 (2014)
13. Onat, N.C., Kucukvar, M., Tatari, O., Egilmez, G.: Integration of system dynamics approach toward deepening and broadening the life cycle sustainability assessment framework: a case for electric vehicles. Int. J. Life Cycle Assess. **21**, 1009–1034 (2016)

Author Index

A
Aboushaqrah, Nour N. M., 279
Acuto, Francesco, 184
Alrawi, Firas, 104

B
Bairamov, Eduard, 238
Bąk, Andrzej, 93
Banet, Krystian, 213
Bogachev, Taras, 126
Bogachev, Victor, 126
Bondarenko, Viacheslav, 114
Burghardt, Tomasz E., 238

C
Çelik, Hüseyin Murat, 37
Chislov, Oleg, 126
Chmielewski, Jacek, 3

D
Dobeš, Peter, 201
Dydkowski, Grzegorz, 13

E
Egorova, Irina, 126

G
Gabsalikhova, Larisa, 153
Galińska, Barbara, 67
Giuffrè, Orazio, 184
Giuffrè, Tullio, 184
Granà, Anna, 184

H
Hadi, Yagoob, 104
Hebel, Katarzyna, 141
Hodás, Stanislav, 201
Hoy, Katarzyna Nosal, 93

I
Ižvolt, Libor, 201

J
Jabbar, Rateb, 279

K
Kłos, Marcin Jacek, 265
Kokoszka, Wanda, 80
Kravets, Alexandra, 126
Kucukvar, Murat, 279

L
Lewicki, Wojciech, 26

M
Macioszek, Elżbieta, 169
Makarova, Irina, 153
Mukhametdinov, Eduard, 153

N
Naumov, Vitalii, 213

O
Olejarz-Wahba, Aleksandra A., 26
Onat, Nuri Cihat, 279

P
Pashkevich, Anton, 238
Pytlowany, Tomasz, 80

R
Radwański, Wojciech, 80

S
Sadygova, Gulnaz, 153
Sendek-Matysiak, Ewelina, 52
Shubenkova, Ksenia, 153
Sierpiński, Grzegorz, 225
Skrzypczak, Izabela, 80
Skurikhin, Dmytro, 114
Sobota, Aleksander, 265
Soczówka, Piotr, 265
Solecka, Katarzyna, 93
Staniek, Marcin, 254
Stankiewicz, Bogusław, 26

Sucha, Matus, 238
Szczuraszek, Tomasz, 3

T
Tóth, János, 225
Tumminello, Maria Luisa, 184
Turoń, Katarzyna, 225

W
Wojciechowski, Jerzy, 114
Wołek, Marcin, 141

Y
Yıldızgöz, Kaan, 37

Z
Zadorozhniy, Vyacheslav, 126
Żochowska, Renata, 265

CPSIA information can be obtained
at www.ICGtesting.com
Printed in the USA
LVHW081042081219
639812LV00007B/1030/P